Confronting Climate Crises through Education

Ecocritical Theory and Practice

Series Editor: Douglas A. Vakoch, METI

Ecocritical Theory and Practice highlights innovative scholarship at the interface of literary/cultural studies and the environment, seeking to foster an ongoing dialogue between academics and environmental activists.

Recent Titles

Ecological Crisis and Cultural Representation in Latin America: Ecocritical Perspectives on Art, Film, and Literature, edited by Mark Anderson and Zelia M. Bora
The Ethics and Rhetoric of Invasion Ecology, edited by Ames Stanescu and Kevin Cummings
Ecotheology in the Humanities: An Interdisciplinary Approach to Understanding the Divine and Nature, edited by Melissa Brotton
*Coexistentialism and the Unbearable Intimacy of Ecological Emergency*edited by Sam Mickey
T.S. Eliot, Poetry, and Earth: The Name of the Lotos Rose, by Etienne Terblanche
Ecofeminism in Dialogue, edited by Douglas A. Vakoch and Sam Mickey
The Image of the River in Latin/o American Literature: Written in the Water, edited by Jeanie Murphy And Elizabeth G. Rivero
Rhetorical Animals: Boundaries of the Human in the Study of Persuasion, edited by Kristian Bjørkdahl and Alex C. Parrish
Ecocriticism in Japan, edited by Hisaaki Wake, Keijiro Suga, and Yuki Masami

Confronting Climate Crises through Education

Reading Our Way Forward

Rebecca L. Young

Foreword by John Adams
Afterword by David W. Orr

LEXINGTON BOOKS
Lanham • Boulder • New York • London

Published by Lexington Books
An imprint of The Rowman & Littlefield Publishing Group, Inc.
4501 Forbes Boulevard, Suite 200, Lanham, Maryland 20706
www.rowman.com

6 Tinworth Street, London SE11 5AL, United Kingdom

British Library Cataloguing in Publication Information Available

Library of Congress Cataloging-in-Publication Data

Library of Congress Control Number: 2018955894

∞™ The paper used in this publication meets the minimum requirements of American National Standard for Information Sciences Permanence of Paper for Printed Library Materials, ANSI/NISO Z39.48-1992.

Printed in the United States of America

Contents

Foreword

Recently I returned from the 2015 Paris COP21 Climate Conference. Along with the effort of diplomats and governmental officials from all over the world to establish policy that would address climate change, there were also seminars on ways to get environmental information out to people throughout the world. Policies are not enough in themselves. Globally, people need to be educated about the crisis we are in, in order to make the necessary changes in human behavior.

The actor Robert Redford, who is a board member of The Natural Resources Defense Council, was interviewed in one of these seminars. He stated that the most important way to communicate is through storytelling. People understand better and are more engaged when they get information through stories.

This is the argument that Rebecca Young is making in her book: *Confronting Climate Crises through Education: Reading Our Way Forward.* Using both classic and contemporary literature, Young shows how literature can serve as a "witness" to social and environmental justice issues. She quotes from known environmental writers such as Bill McKibben and Tom Friedman, but she also shows how such works of classic literature, Mary Shelley's *Frankenstein,* and the contemporary novels of Jonathan Franzen's address environmental and social justice issues. The absence of human companionship and nature create the monster in Frankenstein. Franzen's work, like that of the "muckrakers" in the early days of industrialization in America, is a glaring call to action.

Young makes a strong argument that good literature coupled with environmental science can move us to address immediate environmental issues. She sets a template for ways that literature can be used by educators to bring students to a personal commitment to protect the environment.

John Adams, Founding Director,
Natural Resources Defense Council
January 12, 2016

Acknowledgments

My father and sisters, avid readers who each strive to make the world around them a better place, have nurtured my perceptions of how literature can influence life and individuals can affect change. I am thankful for their natural motivating guidance and to all the family and friends who have supported this project by reading, advising, and encouraging along the way—especially Jennifer Young, who every day embodies the ethics of thoughtful environmental stewardship and informed citizenship.

For providing helpful critiques, alternative perspectives and, most generously, precious time to work, I am very grateful to Michael Gamble.

I could not have completed this book without Marianna Baker and Fran Temple, who graciously saw it through from first brainstorms to final reviews. Their reading suggestions, instructional expertise, and provocative commentary prompted invaluable conversation and reflection at every step of the process. For Fran's logical contributions to planning and Marianna's meticulous editing, I am so appreciative.

The students, colleagues, and mentors I have had the privilege of knowing have inspired hope for the educational reform I outline in this book. The fiction and nonfiction writers who explore not only the environmental crises we are confronting but the potential solutions with which we might overcome them have shown that it is possible.

Introduction

A New Story

"Let us not say there is nothing we could have done. Let this be the beginning of us, rising."[1] Poet Eunice Andrada concludes her performance of "Pacific Salt: A Call for Ocean Protection" with this plea—a call to action that earned her a Spoken Word for the World award in 2015. "Pacific Salt" is a powerful indictment of developed nations' inaction on climate change and a striking catalogue of irreparable consequences for small island nations around the globe.[2] Professor of American and Comparative Literature, Hsin-ya Huang, notes: "In response to climate change and the erosion of their island landscape, the Pacific island writers offer a point of view that dismisses apocalypse and inspires us to imagine adaptation and resilience as our common ecological opportunities."[3] Andrada's is a message of difficult truths, but in its urgency is a vision that demands a new human story—one of hope.

As one of four competition winners invited to the 2015 Paris Climate Conference to share stories of how climate change has affected their lives, then eighteen-year-old Andrada performed her piece at Earth to Paris, a two-day environmental summit complementing the nearby international political conference for climate negotiations. Joining the ranks of many emerging artist activists whose work illuminates climate realities in ways that statistics or data cannot, Andrada represents a rising movement of storytelling to convey personal and social impacts of climate change on cultures all over the world. An earlier event—An Afternoon with Robert Redford: Storytelling for Global Action—was hosted by the United Nations Educational, Scientific, and Cultural Organization (UNESCO) with the purpose of discussing how place-based cultural narratives like Andrada's can influence global action on climate change. Commenting on the event's intention for confronting climate realities Redford said, "I think it takes a lot of voices to create one loud voice to make a change."[4] With "voices" from different parts of the world sharing concerns of indigenous cultures, especially from small island nations, these personal impact stories offer even more compelling reasons for making a change than the scientific data which demand it. Andrada's message is an efficient humanization of this data, her youth a sobering reminder of who ultimately face the impacts of climate change: the collective choice we make now to do nothing or to rise to the challenge will determine how the human story unfolds.

In an interview at UNESCO's event, National Resources Defense Council (NRDC) president Rhea Suh asked Redford about his message for students in the audience. He responded that "we should give the reins to the young people. They are very concerned about the environment. They have the intelligence and the will and the passion to do something about the environment, so let them act. Let them manage what is left on this Earth."[5] As youth representatives like Andrada demonstrate, they are eager to defy predicted outcomes and work toward a better future, yet the task ahead is formidable. Have we prepared them for it?

As a global community we face indisputable climate data and astonishing physical evidence in the form of changing weather patterns and increasing natural disasters. We experience unprecedented levels of pollution, with toxicity affecting air, water, and land. Our economic system drives an unsustainable reliance on dwindling natural resources, while ineffective distribution of resources we do have contributes to crises like poverty, illness, starvation. Forests around the globe, along with the species inhabiting them, are being destroyed for lumber or land profit at a staggering rate. Still more land is irreparably damaged every year for the sake of an increasingly corrupt industrial food system that negatively affects our physical health and moral conscience at every step of the process. Unchecked consumer culture and human population threaten the natural world on multiple levels. Climate-induced migration has and will continue to produce pressure on natural resources as well as political, economic, and social systems. The failure to act responsibly has led to immediate irreversible damage with long-term devastating consequences. As climate scientists have urgently declared, the time for casual action has passed: we must commit ourselves fully to real environmental solutions that support long-range plans for living on this planet in an ecologically appropriate manner.

In the clarifying words of environmental philosopher J. Baird Callicott (and as myriad natural disasters of 2017 portend):

> Global climate change is not something that *might* happen, not just something that surely *will* happen, but something that *is* happening. Glaciers and Arctic

> and Antarctic sea ice *are* melting. Sea levels *are* rising. The oceans *are* acidifying. More frequent and violent storms, floods, and droughts *are* occurring. . . . If we do nothing to slow, eventually to stop, and finally to reverse climate change, the world will be biologically impoverished for the next several million years and *Homo Sapiens* will survive, if at all, in remnant populations living in a state of barbarism and in an interminable Dark Age.[6]

I share Redford's hope that young people have "the intelligence," "the will" and "the passion" to become better stewards of the environment by fighting against this dismal trajectory. What concerns me is that, in my experience as an educator, public education has neither prepared America's youth for climate change challenges nor motivated them toward the career opportunities that will exist in trying to solve them. This book explores the ways education, specifically through literature, can better support students in navigating the uncertain path ahead. What Environmental Humanities Professor in English Gillen D'Arcy Wood argues regarding university education should shape the mission statements of every educational institution:

> Given the truly bleak prospects for our global climatic future, and the urgency of our adapting to it, it is difficult to name a more vital task for the academic humanities than to bring climate change into the classroom. With crisis comes opportunity . . . open[ing] the horizon for a new ecological pedagogy, one that will play a vital role in producing actively informed, climate-progressive students and citizens for the future."[7]

Let us seize this opportunity at all levels of education so that our students might participate in a different human story.

ECOREVOLUTION: LITERATURE FOR A CHANGING WORLD

Environmental educator David W. Orr asserts in *Earth in Mind: On Education, Environment, and the Human Prospect* that while "no generation has ever faced a more daunting agenda" concerning the environment, "we are still educating the young as if there were no planetary emergency."[8] Instead, education's focus seems misaligned with the current climate situation in that—with the exception of some science or special program curricula—academic classes take students further from rather than closer to a connection with the earth. Concluding a chapter titled "The Problem of Education," Orr compares university research on genetic technologies with George Orwell's caution that we might "'reduce the human being to something resembling a brain in a bottle.'"[9] The implications of this research, Orr reflects, will not solve our real needs:

> We need decent communities, good work to do, loving relationships, stable families, the knowledge necessary to restore what we have damaged, and ways to transcend our inherent self-centeredness. Our needs, in short, are those of the spirit; yet, our imagination and creativity are overwhelmingly aimed at things that as often as not degrade spirit and nature.[10]

Throughout, Orr considers how an ecological education can help us better "comprehend our citizenship in the biotic community" by working toward solutions that respect that role and contribute to those needs. Far from being an innocent omission, our failure to do so thus far has contributed significantly to the ecological crises in which we currently find ourselves. In *Ecological Literacy: Education and the Transition to a Postmodern World*, Orr illustrates that our planetary situation demands, without argument, a local, national, and global environmental focus:

> The crisis of sustainability, the fit between humanity and its habitat, is manifest in varying ways and degrees everywhere on earth. It is not only a permanent feature on the public agenda; for all practical purposes it is *the* agenda. No other issue of politics, economics, and public policy will remain unaffected by the crisis of resources, population, climate change, species extinction, acid rain, deforestation, ozone depletion, and soil loss. Sustainability is about the terms and conditions of human survival, and yet we still educate at all levels as if no such crisis existed.[11]

Indeed, there will be no other issue if we do not have a habitable planet. Educating as if there were no crisis invites students to continue the inaction modeled by adults, thus perpetuating a cycle of abuses inflicted on the planet. Effective education can break this cycle.

Orr explains that pedagogy must align with the climate-related problems young people will need to solve. He reminds us that ". . . by what is included or excluded, emphasized or ignored, students learn that they are a part of or apart from the natural world. Through all education we inculcate the ideas of careful stewardship or carelessness."[12] Promoting this stewardship requires thoughtful interdisciplinary instruction through an environmental lens. If students are to understand the relevancy of these connections to their lives, their educators must deliver content in ways that facilitate making those connections. In Orr's words, "Environmental education ought to change the way people live, not just how they talk."[13] As educators, we want students to internalize and subsequently act on their understanding. Thus the preliminary goal of facilitating access to knowledge about environmental crises must lead to the action-oriented goal of nurturing students' commitment to empathy and initiative regarding them.

The general absence of such a focus is not due to a lack of information, for as it relates to current environmental concerns, the global community is rife with scientific data more than sufficient for alarm. Still, despite attempts for reform at the national level, public education in the United States has yet to include environmental literacy as part of any substantial pedagogical or curricular objectives. Within our education system, environmental lessons at the primary, middle, and secondary levels may occur only in science courses or be based on special projects, grants, or teacher initiatives rather than rigorous required curricula. As editors Kathleen Dean Moore and Michael P. Nelson suggest in their introduction to *Moral Ground: Ethical Action for a Planet in Peril*, "clearly, information is not enough. What is missing is the moral imperative, the conviction that assuring our own comfort at terrible cost to the future is not worthy of us as moral beings."[14] I would simply add to this that the information is not coming early enough in our lives to make significant difference. Without a coherent structure for accessing climate-specific data, individually and collectively students are free to rely only on familiar sources that validate what they already believe, or to avoid the subject altogether. Human beings are slow to change, so while it is true that we may have access to plentiful information about environmental crises as well as viable solutions, it is also likely that most of us will be too busy or too habituated to a lifestyle of convenience to heed some randomly encountered fact or guidance, much less seek further information or additional perspectives. The essays collected in *Moral Ground* are meant to reinforce, on both factual and moral grounds, our "obligation to act, and act now."[15] To this end, proponents of ecological education might engage with this collection of multicultural climate change perspectives while focusing on two important questions:

1. How do we provide context to the science so that facts feel personally and culturally relevant?
2. To what extent can we promote a learning environment that transforms apathy or hopelessness about these facts into engaged, prosocial action?

As Orr most recently assesses the state of education in *Dangerous Years: Climate Change, the Long Emergency, and the Way Forward*, the challenge

> is to transform the substance and process of education, beginning with the urgent need to prepare the rising generation—as best we are able—for a rapidly destabilizing ecosphere for which we have no precedent. We cannot know what they will need to know or how they should be taught, but we do know that they will need to see across old boundaries of disciplines, geography, nationality, ethnicity, religion, and time. . . . We must educate them to be the designers of another kind of gyre that turns vicious cycles into virtuous cycles that might someday transform our politics, economy, cities, buildings, infra-

> structure, landscapes, transportation, agriculture, and technologies, as well as our hearts and minds.[16]

To do this, learning must extend beyond the classroom into students' lives—their everyday concerns, decisions, behaviors should reflect what they have understood. From the most basic of agricultural concerns to the more complicated genetic engineering dilemmas, adults have an obligation to educate today's children in ways that prepare them to access and act on information related to climate change. Thus, the questions above are significant not just in individual classrooms or schools but also for the home and workplace. The task, as Orr points out, needs to center on creating a better self in addition to better relationships on familial, communal, national, and international levels. To effectively confront climate change, we need young people to be empowered by education that helps them see the patterns of human behavior that have contributed to it as well as to engage in new patterns that help fight it. We need to foster, as Paul Hawken's work attests, an appreciation of the interconnectedness of social and environmental justice, encouraging more contemplative behavior toward each other, other species, and the environment. *Blessed Unrest: How the Largest Social Movement in History Is Restoring Grace, Justice, and Beauty to the World* explores the magnitude and diversity of a rising movement working to achieve this goal. In the introduction, Hawken explains the connection in a succinct but powerful way:

> A Native American taught me that the division between ecology and human rights was an artificial one, that the environmental and social justice movements addressed two sides of a single larger dilemma. The way we harm the earth affects all people, and how we treat one another is reflected in how we treat the earth.[17]

Like the work of writers included in *Moral Ground*, this speaks to a realization that human beings must reexamine their place in the world. It invites us to consider how we could modify current behaviors—particularly those that will affect future generations—to form a stronger relationship with the earth and each other.

This is precisely why education is so important: to engage young people in perceiving their role in this relationship. Only then will they understand the pivotal role they must take in protecting it. Access to knowledge, debate, and cooperation toward personal and social change helps them do so. At the core of this must be a clear-eyed assessment of our own position accompanied by an honest consideration of the hope we might have for future generations. A sobering start toward this assessment might be to consider emerging data that determine countries' "ecological footprints" with respect to their ecosystem's ability to manage demand in the long term. In "Getting to One-Planet Living" authors Jennie Moore and William E. Rees cite the Global

Footprint Network as part of their argument for an engaged citizenry. Putting our consumption in perspective, Moore and Rees provide this analogy:

> The United Arab Emirates, Qatar, Kuwait, Denmark, and the United States have the largest per capita footprints among countries with populations over 1 million. If everybody consumed like residents of these countries, we would need more than four Earths.[18]

The essay goes on to explore the gross inequities of worldwide ecological footprint data, exposing the chasm between the "wealthy minority" most complicit in ecological consumption and those who will face immediate consequences of this over-consumption, the "more than half the world's population [living] at or below a fair Earth-share."[19] Their review of implications caused by this disparity (driven by a narrative of economic growth) concludes by recommending "a new social contract that recognizes humanity's collective interest in designing a better form of prosperity for a world in which ecological limits are all too apparent and the growing gap between rich and poor is morally unconscionable." As they note, "our individual interests have converged with our collective interests. What more motivation should civil society need to get on with the task at hand?"[20] Their call to action is explicit in its guidance:

> . . . exercise remaining democratic freedoms in the name of societal survival. . . . Ordinary citizens owe it to themselves and the future to engage with their leaders and insist that they begin the national planning process and draft the international accords needed to implement options and choices for an economically secure, ecologically stable, socially just future.[21]

Worldwide progress has been made with the Paris Agreement to address climate change, with several US states and corporations vowing to adhere to its precepts despite the current administration's withdrawal from it. Still, we need "ordinary citizens" to not only demand but embrace such a shift in mind-set. Personal action and public policy must work together to achieve better results in all three areas.

Al Gore's *Our Choice: A Plan to Solve the Climate Crisis* includes a provocative chapter, "Changing the Way We Think," in which Gore asserts the importance of engendering this shift:

> The only meaningful and effective solutions to the climate crisis involve massive changes in human behavior and thinking—changes that lead, in turn, to the widespread use of efficiency and conservation, a shift from fossil fuels to solar and wind and other renewable forms of energy, and an end to the burning of forests and croplands and the depletion of carbon-rich soils.[22]

According to research in neuroscience and studies by behavioral scientists, motivation to change our way of thinking and behaving will come from connecting environmental concerns to "other challenges (economic, strategic, and social) that seem more immediate and are more likely to induce a desire to make the necessary changes."[23] Another significant feature for engagement is how we frame these concerns; it is important, for example, "to link the virtue of solving the crisis to the shared values and aspirational goals that have proved in our past to sustain long-term collective commitments."[24] As social animals, Gore argues, we might be motivated by recognition that cooperation toward environmental solutions is essential to our survival as a species.

Certainly we have an evolutionary history that supports such a conclusion. Hawken's *Blessed Unrest* illustrates in practical form how these suggestions are inherent to a successful global environmental plan. As he explains, today's environmental issues are inextricably linked to the most pressing socioeconomic concerns just as our measure of success may prove directly proportional to the level of collective involvement. The relevancy of our participation is profound as we consider that all species from every single area of the earth are affected.

As we realize the full extent of consequences human beings will face regardless of social position or geographic location, we seek a readily available venue for raising awareness, proposing solutions and mobilizing participation. Education is that venue.

To inspire the broad collective movement Gore and others advocate, two public education reforms must take place concurrently:

1. Infuse and contextualize all curricula with the sense of urgency that climate data illustrate.
2. Address data as patterns of issues that require interdisciplinary solutions.

This is not a movement that can be accomplished through solitary teachers presenting occasional lessons or special projects, but these individuals should be encouraged in their efforts to raise awareness and influence systemic change. Their initiative can motivate parents, administrators, and policy makers to promote the fundamental importance of broadening teachers' and students' environmental awareness while conveying strategies for action. Local issues, for example, are likely to initiate interest, which can lead to collective action plans where groups can be shown how to effect change. As students mature they can draw upon these experiences to extrapolate action plans beyond their immediate communities.

Lest this seem too overwhelming an endeavor, progress on this very goal is being made by a few countries around the world and through the efforts of many thoughtful educators. But, as Thomas L. Friedman suggests in *Hot, Flat, and Crowded: Why We Need a Green Revolution—And How It Can Renew America,* there are significant reasons why the United States should seek a leadership position regarding environmental education. Implications reach far beyond the classroom to impact all aspects of society. In a chapter titled "Green Is the New Red, White, and Blue," Friedman makes the case that solving problems related to climate change is as much an opportunity as it is a challenge.[25] He argues that "'green' is no longer a fad" but is in essence a new currency for the Energy-Climate era we have created:

> . . . the ability to develop clean power and energy-efficient technologies is going to become *the* defining measure of a country's economic standing, environmental health, energy securing, and national security over the next fifty years.[26]

In short, he notes, we are in a "Code Green. It's about national power. Everything America (or any country) can do to go green today will make it stronger, healthier, more secure, more innovative, more competitive, and more respected."[27] America's position as a global power as well as its excessive culpability in current ecological dilemmas are reasons enough to take a leadership role. Fortunately, accepting that role also promises a future of economic, technological, and moral strength. As Friedman frames it—solidifying an argument for all audience interests—America's leadership in the green movement is a win from all angles: "We solve our own problems by helping the world solve its problems. We help the world solve its problems by solving our own problems."[28] As a country we are in an effective position to plan action for a green revolution, to lead the way in new technologies and incentives that help individuals and businesses perform more efficiently as well as more sustainably, to inspire more countries to join the movement and support those already making progress, to make green products and processes more universally accessible. If we view environmental crises from this perspective, the opportunities Friedman's work highlights are clear. Considering the multiple benefits for self-interest and collective success of participating in the sizable global movement already in motion—according to Hawken's research "over one—and maybe even two—million organizations [are] working toward ecological sustainability and social justice"[29] —it seems foolish not to engage more fully.

The public education system makes an effective platform for engaging young people in this movement because it can coordinate multiple levels of involvement including awareness, reflection, activism, and training. Instead, stakeholders are bombarded by a dominant rationale toward economic growth that emphasizes higher science, technology, engineering, and math (STEM) skills for better competition in a global economy. While recent education reforms across the country have emphasized admirable STEM goals, policy makers could realign the process to incorporate our moral compass as well. As Naomi Klein eloquently summarizes in *This Changes Everything: Capitalism vs. The Climate,* our current political focus on economic growth is quite ironically in direct opposition to the needs of the planet's inhabitants:

> Right now, the triumph of market logic, with its ethos of domination and fierce competition, is paralyzing almost all serious efforts to respond to climate change. Cutthroat competition between nations has deadlocked U.N. climate negotiations for decades: rich countries dig in their heels and declare that they won't cut emissions and risk losing their vaulted position in the global hierarchy; poorer countries declare that they won't give up their right to pollute as much as rich countries did on their way to wealth, even if that means deepening a disaster that hurts the poor most of all. For any of this to change, a worldview will need to rise to the fore that sees nature, other nations, and our own neighbors not as adversaries, but as partners in a grand project of mutual reinvention.[30]

Approaching education initiatives from an interdisciplinary perspective that envisions environmental and social justice as primary goals is a worthy endeavor toward this reinvention—one that nurtures intellectual curiosity with ecopsychological empathy. This intentional educative focus could shape a new economic agenda that honors justice as a core value and positions the United States as a contributing leader and conscientious competitor in the Climate Era.

As economist Kate Raworth explores in "Defining a Safe and Just Space for Humanity," the inclusion of information related to environmental and social justice issues is essential to equipping young people with the knowledge and impetus required to advance this position. Suggesting that the "opening page of every macroeconomics textbook" include Johan Rockström and colleagues' instructive diagram of "A Safe and Just Space for Humanity," Raworth proposes:

> So you want to be an economist? Then first, there are a few facts you should know about this planet, how it sustains us, how it responds to excessive pressure from human activity, and how that undermines our own well-being. You should also know about the human rights of its people and about the human, social, and natural resources that it will take to fulfill those. With these fundamental concepts of planetary and social boundaries in place, your task as an economist is clear and crucial: to design economic policies and regulations that help bring humanity into the safe and just space between the boundaries and that enable us all to thrive there.[31]

Providing students with contexts relevant to the interests they pursue is important, but education must also foster significant time, space, and encouragement to discuss these contexts and their future implications. Interested educators and schools might capitalize upon the momentum already established across the nation by forward-thinking institutions to generate such conversations at the elementary, middle, and high school levels. As more colleges and universities implement environmental objectives across their curricular areas we will benefit from the consistency of focus by graduating environmentally and socially conscious young people.

Importantly, this reform should not be perceived or enacted as one more program or standard on top of the complex curricula teachers and students already experience. Instead, it should become the central organizing principle of philosophy and curriculum. Through this lens, schools can adapt existing curricular and building structures to meet educational requirements in all grade level subject areas. At the state level, for example, this means analyzing and delivering standards through a school-wide environmental and social justice philosophy and action plan—one that focuses on local and state issues within a broader context. For schools adhering to the national standards provided in Common Core, the mission statement lends itself to such a focus: "the standards are designed to be robust and relevant to the real world."[32] Regardless of the politics surrounding climate change, there is nothing more relevant than concerns we now face as a local but also global community—from resource shortages to extreme weather to climate refugees. At the international level, we might look to globally conscious education organizations like the International Baccalaureate (IB) to provide unifying reform among schools all over the world. Their mission statement begins with the goal "to create a better and more peaceful world through intercultural understanding and respect."[33] Aspects of their primary years' curricula and overarching philosophy for middle years and diploma levels illustrate this goal as it relates to environmental and social justice. At least anecdotally, students of IB schools seem far more versed in the details of these concerns than students who have not experienced one of the IB programs. A purposeful approach to developing this "understanding and respect" for all students lies in the awareness of how our own cultural beliefs and choices affect others, particularly as focused on the universal need for environmental sustainability.

Under such frameworks, policy makers and educators will be striving toward a shared outcome in preparing students for immediate and future concerns of the world. Regardless of what personal or career paths students choose, this goal is immediately and forever relevant for them as well as for the communities in which they choose to live. Organizing education this way propels us toward affecting individual, social, and global perspectives that can change the direction of environmental practices. More significantly, it will help reestablish humanity's natural and vital connection to the earth and to one another. Informing these perspectives will demand pedagogical paradigm shifts toward ecocriticism and ecopsychology at all levels of education. Activists from all walks of life have started joining forces across the country to implement or enhance education programs that nurture ecological conscience; these grassroots efforts are inspiring the question of how we might implement such change at all levels of education. Models toward such a significant shift can help foster the widescale reform that is needed. This is worth pursuing as we consider the beneficial results achieved by school communities who have experimented with doing so. For public education to interface meaningfully with ecological objectives and governmental policies, redefining goals in the global context of current and future needs is not only essential but inspiring.

Further development of the comprehensive reforms needed are beyond the scope of this single book. In the following chapters, however, I hope to demonstrate that literature is a significant resource for addressing how the transformation might be accomplished. Especially with fiction, students can empathize with voices that speak to environmental and social justice concerns. Exploring how educators can promote such consciousness through a literary lens provides the starting point for more interdisciplinary applications. For example, schools that have already shifted their interdisciplinary program focus from STEM to STEAM (inclusive of the arts), experience an obvious benefit from more arts-based inquiry: creativity. Students who experience learning as interconnected topics rather than specialized subject areas may be better equipped to understand the complexity of climate-related problems and thus to innovate appropriately connected solutions. A thematic climate unit, for example, might include a novel like *Their Eyes Were Watching God*, studied from a lens that focuses on the evolving nature of floods, the impact on lands and people, the role of infrastructure in a changing climate. A science fiction piece like *Do Androids Dream of Electric Sheep* offers opportunities to discuss the impact of species extinction and artificial intelligence on human emotion. These two very different examples indicate the wide range of literature that could enhance traditional STEM programs but there are of course many ways to include other forms of environmentally focused visual and performance arts. As this book will explore, perspectives offered by fiction and nonfiction authors can foster progress at all levels by encouraging students to understand, empathize, and act.

An art that navigates the human condition through history and into our future, literature serves as a platform for environmental pedagogy in all disciplines. Nadine Gordimer's 2001 Nobel Symposia lecture "Witness: The Inward Testimony," explores why we might turn to writers for such a purpose:

> Kafka says the writer sees among ruins 'different (and more) things than others . . . it is a leap out of murderers' row; it is a seeing of what is really taking place.'
>
> This is the nature of witness that writers can, surely must give, have been giving since ancient times, in the awesome responsibility of their endowment with the seventh sense of the imagination. The 'realisation' of what has happened comes from what would seem to deny reality—the transformation of events, motives, emotions, reactions, from the immediacy into the enduring significance that is meaning.[34]

Gordimer addresses the importance of witness literature in the aftermath of September 11th, citing the task of witness writers to define "place," "task," "meaning" in moving forward as a global community. In the Climate Era, we might glean from literature "the enduring significance that is meaning" by engaging with writers who help us understand "place" and "task" as they relate to environment and our interactions within it.

Gordimer's "Witness: The Inward Testimony" is as relevant for global environmental crises as it is for cultural reactions to terrorist violence around the world. The role of writer as witness facilitates readers' interaction with the human experience relating to issues that define, concern, and drive us. By extending our analysis of environmental narratives beyond text, we invite valuable consideration of relevant ethical, political, social, and spiritual contexts. Applying these contexts nurtures more meaningful discourse about humanity's relationship to the natural world, providing a reflective lens by which to evaluate, encourage, or critique behaviors.

An interdisciplinary approach to literature that addresses climate change fosters also an analysis of cultural reactions to mounting evidence—including but not limited to understanding decisions (individual, social, organizational) that lead to environmental and social consequences. Having a firm grasp of human motivations will help determine viable, equitable solutions; similarly, such an understanding could guide what hopefully becomes a moral trajectory for collectively responding to the evidence—evidence that writers who represent diverse backgrounds and interests have sought to "witness" for their readers. While environmental writing is certainly not new, emerging knowledge in the fields of ecocriticism, ecoliteracy, and ecopsychology has contributed to how we engage with it, positioning literature-based ecopedagogy as a key solution to preparing young people for climate crises.

With an ultimate goal of promoting a more environmentally literate student body prepared to confront pressing local and global concerns, the following chapters consider how the way we engage with literature can influence the educational reform we need. Literary selections common to American public education are paired with exemplar instructional models to illustrate accessibility as well as correlation with existing curricula.

In "Literature and Empathy," I argue why literature—a universal art form with a rich history of prompting both individual and social transformative learning—makes an empathic platform for addressing ecological concerns. As an instructive lens, storytelling can be a meaningful resource for initiating reform because it is accessible for all cultures and can be appropriately contextualized by reading level. Its effectiveness lies in authors' ability to reach individuals on a personally engaging level that fosters reflection needed for change.

"A Taker-Leaver Paradigm: Cultural Representations in Contemporary Fiction" extends the discussion by exploring how a selection of literature portrays human's treatment of the earth in ways that may help students understand broader contexts relating to climate change as well as their implications for individual behaviors. Analyses of upper level contemporary fiction that present thematic concerns through a Taker-Leaver paradigm offer individual and cultural representations of these opposing lifestyles—a predominant approach that presumes human exceptionalism against a more marginal perspective that respects balance and sustainability. While the focus is on literary analysis, opportunities for interdisciplinary study include historical, sociological, and psychological contexts.

"Popular Science Fiction and Fantasy: Fostering International Perspectives" examines well-known science fiction, speculative fiction, and fantasy that present environmental and social justice issues in a nonthreatening way. Advocating the need to foster international perspectives in our classrooms, this chapter presents accessible middle, secondary and postsecondary level texts that can help prompt discussion and debate regarding inequitable distribution of natural resources as well as balanced evaluations concerning technological and scientific pursuits. References illustrate the connection between ecological and social concerns, suggesting that the way we respond to characters and conflicts that move us could ultimately help inform actions in our real lives.

Expanding on this, "Let's Share the Table: Building Ecoliterate Communities" promotes food as an essential component of successful ecopedagogical initiatives. This topic introduces many opportunities for traditional literary analysis; in addition to discussions of the interrelationship of characters with Genetically Modified Organisms (GMOS), fair trade issues, and the ethical treatment of farm animals, food in itself is a simple means of bringing together school communities with other stakeholders in the surrounding area. Growing, sharing, and learning about food in our schools invites community members from various backgrounds—farming, architecture, medicine, manufacturing, waste management—to participate in the education of young people and inspire solution-based career interests and goals.

"Morality and Environmental Responsibility: An Interdisciplinary Reading of Franzen's *Freedom*" employs an interdisciplinary lens that applies

Patricia Churchland's *Braintrust*—on the neuroscience of human morality—to a literary analysis of Johnathon Franzen's novel *Freedom*, which recounts the dangers of human exceptionalism and overpopulation, species extinction, and depletion of natural resources through the dynamics of an American family struggling to love and forgive one another. It is a more focused secondary or post-secondary example of how we can merge the humanities and sciences, an approach encouraged throughout this book.

Finally, "Ecopsychology: Harmonizing Our Paths" illustrates individual and social benefits of an education system that promotes understanding of the world around us as a means of better understanding ourselves. It includes a discussion of specific ecopedagogical practices at various levels of education. Fiction examples consider the ways humans interact with each other and the natural world through symbolic representations offered by iconic characters, echoing aspects of individual and social morality referenced in the previous chapter. The argument urges our return to an evolutionary ecopsychological connection to Earth over a posthumanist solution for confronting climate change.

One need not be a scientist or politician or activist to recognize that we all benefit from a future of healthier, more sustainable interactions that seek to mitigate the effects of climate change. I am not suggesting that education alone can or will act quickly enough to do so; the climate crisis we face demands far more immediate action than long-term pedagogical goals can achieve. However, ecologically appropriate social consciousness can be encouraged so that taking action on climate change is not so mired in controversy, fear, and denial for next generations but instead becomes natural participation in informed environmental and social justice initiatives. As described in *Teaching Climate Change in the Humanities,*

> [t]eaching can be a kind of weaving, a stitching together of communities both within and beyond the classroom. Teaching climate change in the humanities, with attention to representations and mediations, histories and ethics, material artifacts and cultural forms, can push us through despair, perhaps, and toward the social practice of caring—a practice not identical to hope but at the least embedded in relationships that compel us to place questions of justice and of collective survival at the forefront of our thinking.[35]

This effort should and must be the energizing moral force of our time. Literature can help.

NOTES

1. Eunice Andrada, "Pacific Salt: A Call for Ocean Protection," *YouTube*, 28 Nov. 2015, https://www.sbs.com.au/yourlanguage/filipino/en/audiotrack/pacific-salt-call-ocean-protection. Accessed 7 Nov. 2017.

2. Caroline Saunders, "The Beginning of Us, Rising—Storytelling for Climate Change," *CONENERGY*, 8 Dec. 2015, conergy.com/de/node/231. Accessed 10 Jan. 2017.

3. Hsinya Huang, "When Sea Levels Rise: Writing/Righting Climate Change in Pacific Islanders' Literature," in *Teaching Climate Change in the Humanities* (New York: Routledge, 2017), 263.

4. "Robert Redford: Give the Reins to the Young," *UNESCO*, 7 Dec. 2015, //en.unesco.org/news/robert-redford-give-reins-young. Accessed 10 Jan. 2017.

5. Ibid.

6. J. Baird Callicott, "Thinking Climate Change Like a Planet: Notes from an Environmental Professor," in *Teaching Climate Change in the Humanities* (New York: Routledge, 2017), 8–2.

7. Gillen D'Arcy Wood, "Teaching Teleconnection," in *Teaching Climate Change in the Humanities* (New York: Routledge, 2017), 96.

8. David W. Orr, *Earth in Mind: On Education, Environment, and the Human Prospect* (Washington, DC: First Island Press, 2004), 27.

9. George Orwell, *The Road to Wigan Pier* (New York: Harcourt Brace Jovanovich, 1958), 201, quoted in Orr, *Earth in Mind,* 33.

10. David W. Orr, *Earth in Mind,* 33.

11. David. W. Orr, *Ecological Literacy: Education and the Transition to a Postmodern World* (Albany: State University of New York Press, 1992), 83.

12. Ibid., 90.

13. Ibid., 91.

14. Kathleen Dean Moore and Michael P. Nelson, "Toward a Global Consensus for Ethical Action," in *Moral Ground: Ethical Action for a Planet in Peril* (San Antonio, TX: Trinity University Press, 2010), xvi.

15. Ibid., xviii.

16. David W. Orr, *Dangerous Years: Climate Change, the Long Emergency, and the Way Forward* (New Haven: Yale University Press, 2016), 10–06.

17. Paul Hawken, *Blessed Unrest: How the Largest Social Movement in History Is Restoring Grace, Justice, and Beauty to the World* (New York: Penguin Group, 2007), 2.

18. Jennie Moore and William E. Rees, "Getting to One-Planet Living," in *State of the World 2013: Is Sustainability Still Possible?* (Washington, D.C.: Worldwatch Institute, 2013), 41.

19. Ibid., 43.

20. Ibid., 49.

21. Ibid., 49-50.

22. Al Gore, *Our Choice: A Plan to Solve the Climate Crisis* (New York: Rodale Inc., 2009), 315.

23. Ibid.

24. Ibid.

25. Thomas L. Friedman, *Hot, Flat, and Crowded: Why We Need a Green Revolution—And How It Can Renew America* (New York: Farrar, Straus and Giroux, 2008), 171.

26. Ibid., 172.

27. Ibid., 173.

28. Ibid., 173.

29. Paul Hawken, *Blessed Unrest* (New York: Penguin Group, 2007), 2.

30. Naomi Klein, *This Changes Everything: Capitalism vs. The Climate* (New York: Simon & Schuster, 2014), 23.

31. Kate Raworth, "Defining a Safe and Just Space for Humanity," in *State of the World 2013: Is Sustainability Still Possible?* (Washington, D.C.: Worldwatch Institute, 2013), 35.

32. Common Core State Standards Initiative, "Mission Statement," Common Core State Standards Initiative: Preparing America's Students for College and Career. 2012. http://www.corestandards.org/. Accessed 12. Nov. 2016.

33. "Mission and Strategy," *International Baccalaureate Organization.* http://www.ibo.org. Accessed 12. Nov. 2016.

34. Nadine Gordimer, "Witness: The Inward Testimony," *Nobel Centennial Symposia,* www.nobelprize.org.

35. Stephen Siperstein, Shane Hall, and Stephanie LeMenager, *Teaching Climate Change in the Humanities* (New York: Routledge, 2017), 9.

Chapter One

Literature and Empathy

A Rationale for Change

As avid readers know, literature can prompt self-reflection—perhaps leading to actions that unemotional data may not. Its universality implies an art form with far-reaching potential for inspiring change. If in our reading lives we engage with characters that bring clarity through narrative about ecological and social issues, we might find those concerns resonating in our lives, demanding not only interest but action. This chapter explores why literature makes an effective resource for introducing environmental and social justice concerns in the classroom.

Novelist and Cognitive Psychologist Keith Oatley reviews in *Theory of Mind and Theory of Minds in Literature* how cognitive cultural and literary studies illuminate literature as a platform for powerful engagement. In an analogy Oatley applies to literature's potential in helping us understand others and ourselves, he asserts, "[h]aving a Theory of Mind is like having a watch or a clock. These devices are models of the rotation of the earth. The positions of sun and stars are indications of time, but they are difficult to judge accurately, and heavenly bodies are not always visible. A model—a watch or clock—is much better."[1] As Oatley demonstrates in the literary example he employs, the practice of mind reading we engage in when we encounter others (including characters of literature) may help us better comprehend—much like a watch or clock allows us to more accurately pinpoint time—future interactions we have. As he explains, "the idea of Theory of Mind refers to knowing what another person may be thinking or feeling in the moment, but with people whom we are likely to interact with on future occasions, we gather information to build mental models of them over the long term, from their utterances, from their behavior, and from what other people say about them."[2] The benefit of this analysis we engage in over time has the potential to foster better relationships as we learn to understand the motivating characteristics of those around us; the more we practice this skill through reading, the more perceptive of others we may become.

Professor Patrick Colm Hogan explores a similar perspective focused on the universality of verbal art, commenting that it is "in fact highly surprising that verbal art is produced in small nomadic groups as well as in vast, highly urbanized nations" since "there is no logical necessity in [its] existence."[3] Our most common forms of literature—prose, drama, poetry—are similarly universal, suggesting that:

> Tale telling [also] appears to be a literary universal. Probably in all societies, people articulate causal sequences of nonbanal events involving human agency (with banality defined relative to culturally specific expectations), and do so at least in part for aesthetic enjoyment, itself based on identification, the patterned variation of emotional intensity, and so forth.[4]

As Hogan articulates, societies also share particular aspects regarding the types of stories told—on "love and political power,"[5] for example, and reveal further similarities in the ways we interact with fictional characters: "It appears that readers and auditors identify with a character on the basis of shared high-leveled properties in their self-conception and that they prefer works involving characters with whom they identify."[6] While the properties by which readers relate to characters will vary by culture and indeed by individual, we can see how these literary connections could be important in navigating environmental issues by interacting with each other and other countries to solve them.

In addition to identifying with fictional characters, evidence suggests that we might learn more about our problems (and, by extension, ourselves) through them. Independent scholar Alan Palmer discusses cognitive approaches to literary study in his essay "Storyworlds and Groups," suggesting that since "we all think of novels in terms of the mental functioning of characters" readers are "all cognitivists."[7] One's reading skills, therefore, might benefit from understanding the real motivations and choices of others as well as our own. Palmer summarizes the work of social psychologist Timothy Wilson to assert why this skill is important:

> . . . our private thought is not immediately accessible and available to us. *We have to infer what we ourselves are thinking in much the same way as we infer what other people are thinking.* We deduce the nature of the unconscious mind by looking outward at the behavior that results from it.[8]

Thus, in the case of literature, a study of characters' thoughts and actions might provide interesting insight about our own. Lisa Zunshine's work on

Theory of Mind suggests that "our tendency to explain observed behavior in terms of underlying mental states seems to be so effortless and automatic because our evolved cognitive architecture 'prods' us to learn and practice mind reading daily, from the beginning of awareness."[9] As Zunshine notes in *Why We Read Fiction: Theory of Mind and the Novel*, "from a cognitive perspective, a misinterpretation of a character's state of mind is still very much an interpretation, a fully realized and thus pleasurable engagement of our Theory of Mind."[10] In fiction as in life individuals may not interpret perceptions with complete accuracy, but the process itself illustrates how interactions with characters, conflicts, and settings can inform the way we understand and respond to real people, problems, and places.

Hogan's explanation of why readers respond emotionally to literature, despite a cognizance of it being fictional, further suggests its transformative potential:

> . . . in a neurobiological sensitivity model, the fictionality of the work is irrelevant. . . . One function of emotion for humans, and one important reason for the involvement of the prefrontal cortex in emotion circuits, is to guide our selection of future behaviors.[11]

As Hogan points out, since "emotion is a crucial part of the function of imagination," it is natural for us "to expect emotional responses to fiction."[12] Further, one's response to fiction will be in proportion to the concreteness of the story presented. In the case of literature about the environment, for example, readers may be more inclined to take notice of issues effectively described in fiction that relates to their lives than of actual data that may not.

Because emotional responses contribute to the cognitive processing of social, political, moral, or ethical issues presented, engagement with fiction might facilitate learning. Philosopher Noël Carroll's comparison of literature to the thought experiments of philosophy helps explicate his assertion that art and literature can indeed produce a learning experience for viewers or readers. His essay "The Wheel of Virtue: Art, Literature, and Moral Knowledge" explores the nature of philosophy's thought experiments, which are often fiction and sometimes derived from fiction literature. As he notes, thought experiments make

> connections—that were hitherto recessive or obscure—between what is already known and other parts of our cognitive stock. They illuminate the relevance of what is already known to the question at hand by refocusing that knowledge in a novel way. This counts as knowledge production, because it clarifies linkages between parts of our cognitive map.[13]

By extension, the same concepts might be applied to literature, which "enters public discourse at the level of ordinary experience"[14] and reaches a more general and broader audience than authoritative morality philosophers, for example. Importantly, authors may challenge the accepted ideas or practices of a place or culture through their fictions:[15]

> Art and literature, by destabilizing reigning ideas, often elicit social debate by functioning as counter-examples, undercutting, subverting, and disconfirming common belief by designing thought experiments. . . . Much artistic criticism of social convention operates like this: through imagining eminently possible cases that contradict conventional wisdom, authors prompt readers to recognize the limitations of the maxims and concepts they live by as well as their implications.[16]

As in Plato's Allegory of the Cave, what students choose to do with this information will be up to them but we can hope they choose to confront it.

Exploring E. M. Forster's novel *Howard's End* as an exemplar of "literary and artistic thought experiments [that] [concern] our concepts of virtue and the conditions according to which we apply them,"[17] Carroll explains the novel's structure (which he suggests is common in narratives of art and literature) as indicative of the thought and discussion it might promote:

> A virtue wheel or virtue tableau comprises a studied array of characters who both correspond and contrast with each other along the dimension of a certain virtue or package of virtues—where some of the characters possess the virtue in question, or nearly so, or part of it, while others possess the virtue, but only defectively, or not at all, even to such an extent that their lack of the virtue in question amounts to the vice that corresponds to the virtue.[18]

Noting another classic's virtue wheel structure, Carroll frames Charles Dickens's *Great Expectations* as a fiction that "enables us simultaneously to clarify our concept of virtuous parenthood—it allows us to reflect upon the conditions where we would apply or withhold it—while also sharpening our sense of when the concept is defectively or even viciously instantiated."[19] The effect of such textual structures is such that they "make these conceptual discoveries available to us, if not as we read along, then in the reflective afterlife of the text or in conversation with others about our reaction to the text."[20] Given the various topics available in our literary lives, it is easy to see what broad educative value literature presents—both in the classroom and outside—reading for pleasure, enjoying a play or a film. This "reflective afterlife" is especially cherished by literature instructors and is the ideal period to offer students opportunities to discuss and act on their reflections—promoting a class community project or local activism or volunteerism, for example.

Because of the variety and richness of contexts literature can provide, its instructional potential regarding moral and ethical considerations is extensive. Learning may take place in responses and discussions of literary experiences because readers "understand these characters in terms of virtually the same person schemas [we] use to understand each other in everyday life":[21]

> As we compare and contrast what we have observed about the characters with the views of others, and as we weigh our judgments against what others have to say, we are, of course, engaging in a process of reflective equilibrium about the aptness of our applications of concepts of virtue and vice that, among other things, brings to the fore or makes available to use an awareness of the conditions in accordance with which we make those applications and issue our judgments.[22]

In my own experience as a teacher, fiction that has provided such engaging and illustrative social and environmental reflection include Mark Twain's *Huckleberry Finn*, Harry Mulisch's *The Assault*, Margaret Atwood's *The MaddAddam Trilogy*, Shakespeare's *A Midsummer's Night Dream*, Nadine Gordimer's *The Pickup,* Cormac McCarthy's *The Road,* Isabelle Allende's *House of the Spirits,* Mary Shelley's *Frankenstein*, George Orwell's *1984*, and Paolo Bacigalupi's *The Windup Girl* to name a few. The variety of issues each piece presents invariably leads to complex questions and discussions, ultimately offering readers a premise or context for considering how they feel and think about moral, political, social, or personal situations that characters face, even those they face themselves.

This contextualization could benefit studies of environmental and social justice literature, which rely on an appreciation of often complex background information. In "Narrative, Imagination, and the Search for Intelligibility in Environmental Ethics," Philosophy Professor Roger J. H. King offers the following explanation regarding why this approach is effective for facilitating learning about environmental issues:

> Given the structure a particular story provides for the interpretation of human experience, the environmental philosopher must choose whether to construct arguments within the confines of the ongoing narrative expectations or set out to write a new narrative. . . . These fictional narratives may take many forms: the narratives of nature writers, utopian and dystopian novels and films, landscape paintings, or moral philosophy. From the contextualist point of view, it makes sense to include moral philosophy in the list of fictional narratives, because the use of narrative and imaginative resources is 'internal' to the practice of ethics, not merely 'external' and illustrative.[23]

If the way we perceive morality and contemplate our own environmental and social choices is dependent on the way we view ourselves in the human story, then we should ensure students of literature have narrative models by which to ethically evaluate their roles. As King deduces from philosopher Alasdair McIntyre's work, "moral choice presupposes that we are able to situate ourselves, see our place in a narrative structure, and see how our alternatives contribute to the action that is always already ongoing."[24] Thus, in the case he describes regarding efforts to create appreciation for wilderness preservation, he notes the importance of "embedding our understanding of holistic value in narratives of human life that flesh out visions of how humans should interact with nonhuman nature."[25] Particularly in the case of the millennial generation, for example, literature could supplement the nature narrative that seems absent from their everyday lives. It invites response to an alternative narrative (one that is not human-centered) regarding their relationship to nonhuman species and to the natural environment. How readers emotionally react to such narratives could inform how they proceed to act within their actual environments. Alternatively, children will continue to be more readily able to identify symbols of mass consumerism and technology than those of nature, as demonstrated in the Ground Productions documentary *Play Again.* The film, which "investigates the consequences of a childhood removed from nature and encourages action for a sustainable future"[26] follows six teenagers who are asked to leave their technology aside for an experience in the outdoors. It demonstrates a hopeful message that young people who experience outdoor play and the beauty of nature may learn the importance of engaging with and protecting it.

Professor of English Suzanne Keen's "A Theory of Narrative Empathy" establishes her larger goal of exploring whether Theory of Mind and responses to literature that evoke empathy lead to real action. Her work examines "the literary version of the empathy-altruism hypothesis, which holds that novel reading, by eliciting empathy, encourages prosocial action and good world citizenship."[27] According to Keen's research,

> readers' perception of a text's fictionality plays a role in subsequent empathetic response, by releasing readers from the obligations of self-protection through skepticism and suspicion. Thus they may respond with greater empathy to an unreal situation and characters because of the protective fictionality, but still internalize the experience of empathy with possible later real-world responsiveness to others' needs.[28]

Potential research for linking narrative empathy to real-world application or action presents stimulating questions for educators who seek to enhance the learning experiences of students—particularly regarding social justice issues. Studying and teaching fiction, for example, may be enhanced by exploring the findings of cognitive scientists on mirror neurons ("neurons that discharge both when an individual executes a motor act and when he observes another individual performing the same or a similar motor act"[29]) and affect. Keen's essay includes compelling evidence why we should pursue a better understanding of this link between literature and empathy education:

> The possibility that novel reading stimulates mirror neurons' activation can now, as never before, undergo neuroscientific investigation. Neuroscientists have already declared that people scoring high on empathy tests have especially busy mirror neuron systems in their brains. Fiction writers are likely to be among these high empathy individuals. For the first time we might investigate whether human differences in mirror neuron activity can be altered by exposure to art, to teaching, to literature.[30]

Considering how we might employ literature as an avenue for environmental and social justice discussions, teachers could select fiction emphasizing techniques that contribute to reader empathy. As Keen observes,

> Narrative theorists, novel critics, and reading specialists have already singled out a small set of narrative techniques—such as the use of first person narration and the interior representations of characters' consciousness and emotional states—as devices supporting character identification, contributing to empathic experiences, opening reader's minds to others, changing attitudes, and even predisposing readers to altruism.[31]

What literature teacher or reader wouldn't feel empowered by such results? The science here may only reaffirm what good writers and readers already believe, but the data validate educators' fiction selections and pedagogical choices—particularly under the pressure of Common Core's focus on informational texts over fiction. However, as Keen cautions in the preface of *Empathy and the Novel*, "we have a great deal more to understand about narrative empathy before we place the novel in service to social goals" and there is a broader educational context to consider:

> Books can't make change by themselves. . . . If, however, a society judges that its peoples' novel reading ought to invoke empathy that yields in committed action for others, then it should be willing to invest in the teachers who will help readers make connections between their feeling responses to fiction and subsequent behavior as citizens of the world.[32]

This is a statement most educators would support. Such a goal for transformative experiences prompted by literature is important to consider in assessing curricular objectives related to environmental and social justice. I do not suggest we employ literature merely as a tool for achieving these objectives. Students also benefit from traditional literary analyses by which they become more mature readers able to engage with more sophisticated texts. Rather, judicious choices of fiction studied in classrooms around the world could provide an effective lens for studying and discussing issues of enormous magnitude that educators might otherwise avoid—for lack of administrative support, limits to our own knowledge, or fear of negative reactions. Fiction provides a safe space to examine controversial topics and to assess students' readiness or interest in learning more about them.

This defense of fiction of course extends beyond the classroom as well. In a thoughtful lecture, "Why our future depends on libraries, reading, and daydreaming," writer Neil Gaiman makes such a case with his emphasis on saving libraries and fostering a love for reading and imagination. Arguing the importance of reading fiction for pleasure, Gaiman suggests its role in promoting further reading and writing and in nurturing empathy—key aspects for the betterment of communities:

> You get to feel things, visit places and worlds you would never otherwise know. You learn that everyone else out there is a me, as well. You're being someone else, and when you return to your own world, you're going to be slightly changed.[33]

When writers and readers learn how to maximize this experience, change we undergo as artists and audiences of fiction might have the potential to be transformative in our approach to other humans, other species, and our environments. The ability to empathize with characters and places in fiction may help us better understand ourselves and our interactions in the world. Individually empowering but also socially significant, Gaiman asserts that fiction represents the possibilities of imagination which can lead to new ideas, problem-solving, and innovation in our real lives:

> Once you've visited other worlds, like those who ate fairy fruit, you can never be entirely content with the world that you grew up in. Discontent is a good thing: discontented people can modify and improve their worlds, leave them better, leave them different.[34]

We want to encourage the kind of reading that leads to these responses because resulting interactions can be so powerful. Readers who ask "why" and "what if" are those who will implement needed changes in their communities and world.

In "Teaching Ethical Know-How in New Literary Spaces," Rebecca Luce-Kapler, Dennis Sumara, and Tammy Iftody explore how these interactions occur in both physical and virtual narratives. Regarding fictional identifications, they discuss mind reading or Theory of Mind as a "biological capacity" that allows us to "become more deeply aware of the contours of our own consciousness by virtue of opportunities to interpret and imagine the consciousness of others,"[35] noting significantly the connection between this capacity and our reading skills:

> . . . new studies in consciousness have important links to literary processes. Donald (2001), Lodge (2002), and Zunshine (2006) suggested that the literary

> novel is the symbolic cultural form par excellence for capturing the phenomenological intricacies of human mentality (or consciousness) and is therefore a rich resource for developing more sophisticated (empathic) mind-reading abilities.[36]

How we might teach our students empathetic reading strategies is a concern they address, inferring from "what Varela (1999) called 'ethical know-how'" that "ethics is not a set of a priori principles that can be applied to any situation; instead, ethics emerge from our empathetic identifications with others."[37] Encouraging close-readings of texts that "[teach] our students the power of imagining minds along with the ethical imperative" will challenge readers to consider alternative narratives, to "imagine the mind of another."[38] Not surprising to instructors of literature or anyone who has observed a reader's deep connection with a character or story, empathetic reading has value in addressing myriad social and environmental concerns and is further evidence of the importance of literature in societies. In "Social Anthropology and Literary Sensibility," M. N. Srinivas interprets sociologist Marcel Mauss' observation that "'the sociologist has to have the sensitivity of a novelist.'"[39] As Srinivas explains, "understanding a different culture is essential to understanding one's own. It is a truism that the study of 'the other' enables one to understand oneself."[40] Thus, as readers we might take a similar anthropological approach to learning about and appreciating other cultures; the literature of a place "provides a window to the thoughts, emotions, values, dilemmas and conflicts"[41] of its people. As a source of such information, the study of literature might serve humanity in other cultural areas such as science, language, history, and the arts.

Through an environmental lens, literature can provide cultural perspectives that help us situate particular ecological issues within the causal variables, consequences, and achievements of a specific place and its people. As Humanities and Philosophy Professor Stephen Davies offers in "Responding Emotionally to Fictions," it is imperative that we consider how narratives affect not only our lives but also the overall histories and contexts of them.[42] He suggests that narratives "feature hypothetical and counterfactual scenarios as part of making sense of what did happen, what is happening, and where we will go from here."[43] Since "we respond emotionally to scenarios that are hypothetical or counterfactual as readily as to ones that are actual," the "affective engagement that narratives demand from us" is as significant in our "quest for fictional entertainment as in factual, quotidian circumstances."[44] In exploring the paradox of readers' responses to fictional versus real events, Davies offers as example an account of his feelings toward real-world war scenarios. His sympathetic response to Dickens' young protagonist Oliver Twist suggests that a reader's capacity for empathy—once awakened—might be transferred to other social contexts:

> . . . It might be suggested that, if my feeling is genuine, I should act to help others avoid the same fate. My pity of the victims of war should lead me to question the point and legitimacy of modern-day wars, for instance. But if this is true, the same applies to my response to the fictional character. I cannot change Oliver's world, but the story should lead me to oppose real-world policies that tolerate poverty and child labor.[45]

Thus, having read *Oliver Twist* and discussed effects of hunger, poverty, or urban crime in that era, the reader might read about other topics to make similar extrapolations about war and the impacts of current and future armed conflicts. Some readers will need more modeled discussion and guidance, but if we read with this focus—beginning early and practiced throughout middle and upper grade levels—we could graduate more civil-minded students.

Davies' example further highlights a key aspect regarding the hesitation to act on issues we become aware of through fact or fiction. As he asserts, "being moved to action by one's emotions" "is likely only where one's actions can make a relevant difference."[46] Showing students how their actions can make an impact—either positive or negative—must therefore be part of the study, which may seem a more difficult task than it is: "Undoubtedly we should be moved by the plight of real people, but it is not obvious that we should always be moved less or not at all by fictional characters, who are often more vividly present to us than anonymous real-world sufferers."[47] In fact, as Davies candidly points out, we are even likely to be more moved by the characters we engage with than by real people whom we've never met. Thus, with respect to studying issues of concern through the lens of fiction, readers might be compelled to act in ways that are relevant and accessible in their real lives. Fiction that addresses environmental issues, for example, might encourage us to change the way we think and behave since ecological sustainability is an obviously universal theme. Feeling empathy for and connecting with a character who laments the loss of rain forests, for example, may not move one to launch a campaign against deforestation, but it could motivate one to be more discerning in the choice of wood furniture.

Professor of Ecology F. Stuart Chapin III explains why literature study might be such an important strategy in addressing environmental dilemmas such as climate change in his essay "The Planet Is Shouting but Nobody Listens":

> The scientific community has been relatively ineffective in conveying this message of planetary change to our society, whose collective choices propel us along this path. As scientists, we are trained to avoid speaking in ways that touch people's souls. New forms and venues of dialogue must emerge through words, images, and tones that enable society to see and hear more clearly the changes that are occurring, to feel deeply and personally their importance, and

> to recognize the connections between our personal and collective choices and the trajectory of life on this planet.[48]

To his point, this seems precisely what novelists like Barbara Kingsolver are trying to inspire through their climate fiction or cli-fi, a field which is quickly gaining popularity with readers' increasing consciousness toward the environment. In *Flight Behavior*, Kingsolver presents a mother coming to grips with the consequences of climate change in a personal way as the Tennessee farm her family lives on has become the unlikely stop for millions of migrating monarch butterflies. Despite having no formal training, Dellarobia has started a job assisting the scientist who is staying on her property while his team studies "why a major portion of the monarch population that has overwintered in Mexico [. . .] would instead aggregate in the southern Appalachians, for the first time in recorded history."[49] Dellarobia has limited awareness or knowledge about climate change, a term which she is cautious about mentioning around her family. Her observations, however, of the scientists' concern as they desperately try to collect evidence on the butterflies before the January cold "lull[s] [them] into a dormancy from which they would not wake"[50] turn to concern about the dreams of her young son and the future reality he may face:

> With all his might Preston wanted to be a scientist and study animals. But in the lab Dellarobia listened to Ovid and Pete speaking hopelessly about so many things. The elephants in drought-stricken Africa, the polar bears on the melting ice, were 'as good as gone,' they said with infuriating resignation as they worked through what seemed to be an early autopsy on another doomed creature. *Gone*, as if those elephants on the sun-bleached plain were merely slogging out the last leg of a tired journey. The final stages of grief. Dellarobia felt an entirely new form of panic as she watched her son love nature so expectantly, wondering if he might be racing toward a future like some complicated sand castle that was crumbling under the tide. She didn't know how scientists bore such knowledge.[51]

This passage illustrates what Chapin suggests is lacking in the presentation of available scientific data: "the missing pieces are the dialogues necessary to connect the increasingly obvious planetary changes with the deepest motivations of every person as a steward of planet Earth."[52] Like Dellarobia, readers may have not personally confronted or come to terms with issues related to climate change, yet this mother's fear about the future her son may face as a result of indisputable facts and unstoppable trajectories—however unaware of them she may be—is in many ways more persuasive to the average reader than are the actual statistics. Readers can identify with her concern and thus perhaps more easily relate to her protesting the planned logging of trees the butterflies now inhabit—despite the conflict it creates for her, her husband,

and father-in-law, who has already contracted with a logging company to help repay his loans. Kingsolver develops this dilemma so that through Dellarobia's new stewardship of the trees and her own recognition of the difficult choices we sometimes face, it is not easy for readers to judge her father-in-law. Through this character, we quickly learn that the underlying causes of climate change are not so clearly defined by "right or wrong," the people involved not so easily categorized as "good or bad." What Kingsolver's storytelling provides is the opportunity for readers to delve into gray areas seeking equitable solutions that honor the places, people, and other species involved.

Professor of Law Robin Morris Collin suggests the important outcomes that might come of such discussion and preparation in her environmental essay "Restoration and Redemption":

> Our power to choose is both small and vast. It ranges from choices about what to eat to decisions to design whole systems free of waste and pollution. Perhaps the most powerful exercise of our powers lies in our abilities to create stories to frame our sense of reverence and to tell those stories to you, our children.[53]

These choices—from the groceries we buy to the policies we support or don't—contribute to a collective response to climate change that will determine our future and the story we tell. The universality of literature provides an awesome opportunity for reaching vast, global audiences and fostering important conversations that inspire more informed actions. Part of the challenge, for educators, beyond encouraging new perspectives on what it means to live responsibly on the earth is preparing students to become better stewards. As Professor of English William Gleason asserts in "Looking Back to Look Ahead: Climate Change and US Literary History,"

> we would serve ourselves well by bringing to the foreground—wherever we can—those texts and writers that, in different periods and genres, have engaged with the unsettling effects of environmental change. After all, the more ways we can help our students develop a surer sense of the power of cultural forms to engage "real-world" challenges, the better prepared we will make them to grapple with the future of climate change, which is already as pressing a cultural question as it is a scientific one.[54]

Empowering young people to make better choices in the face of these challenges than have their predecessors is the way we begin to change our climate story, to become "the story of us, rising."

NOTES

1. Keith Oatley, "Theory of Mind and Theories of Minds in Literature," in *Theory of Mind and Literature*, eds. Paula Leverage, Howard Mancing, Richard Schweickert, Jennifer Marston William (Purdue University Press, 2011), 14.
2. Ibid., 16.
3. Patrick Colm Hogan, "Literary Universals," in *Introduction to Cognitive Cultural Studies*, ed. Lisa Zunshine (Baltimore: The Johns Hopkins University Press, 2010), 4–5.
4. Ibid., 45.
5. Ibid.
6. Ibid., 50.
7. Alan Palmer, "Storyworlds and Groups," in *Introduction to Cognitive Cultural Studies*, 188.
8. Ibid.
9. Lisa Zunshine, "Theory of Mind and Experimental Representations of Fictional Consciousness," in *Introduction to Cognitive Cultural Studies*, 196.
10. Lisa Zunshine, *Why We Read Fiction: Theory of Mind and the Novel* (Ohio State University Press, 2006), 25.
11. Patrick Colm Hogan, "On Being Moved: Cognition and Emotion in Literature and Film," in *Introduction to Cognitive Cultural Studies*, 246.
12. Ibid.
13. Noël Carroll, "The Wheel of Virtue: Art, Literature, and Moral Knowledge," *The Journal of Aesthetics and Art Criticism* 60, no. 1 (Winter 2002): 8, http://www.jstor.org/stable/1519970.
14. Ibid, 10.
15. Ibid.
16. Ibid., 11.
17. Ibid., 11.
18. Ibid., 12.
19. Ibid., 12.
20. Ibid., 12.
21. Ibid., 15.
22. Ibid.
23. Roger J.H. King, "Narrative, Imagination, and the Search for Intelligibility in Environmental Ethics," *Ethics and the Environment*, 4, no. 1 (1999): 30, http://www.jstor.org/stable/40338957.
24. Ibid, 29.
25. Ibid.
26. *Play Again*, Ground Productions, accessed March 15, 2017, http://playagainfilm.com/film-synopsis/.
27. Suzanne Keen, "A Theory of Narrative Empathy," *Narrative* 4, no. 3 (October 2006): 224, http://www.jstor.org/stable/20107388 .
28. Ibid., 220.
29. Sourya Acharya and Samarth Shukla, "Mirror Neurons: Enigma of the Metaphysical Modular Brain," *Journal of Natural Science, Biology, and Medicine* 3 (2) (2012 Jul–Dec): 118, doi: 10.4103/0976-9668.101878.
30. Suzanne Keen, "A Theory of Narrative Empathy," 207.
31. Ibid., 213.
32. Suzanne Keen, *Empathy and the Novel* (Oxford University Press, 2007), xv.
33. Neil Gaiman, "Why our future depends on libraries, reading, and daydreaming," *The Guardian*, October 15, 2013, http://www.theguardian.com/us
34. Ibid.
35. Rebecca Luce-Kapler, Dennis Sumara, and Tammy Iftody, "Teaching Ethical Know-How in New Literary Spaces," *Journal of Adolescent and Adult Literacy* 53, no. 7 (April 2010): 538, http://www.jstor.org/stable/25653903.
36. Ibid.

37. Ibid, 539-540.
38. Ibid, 540.
39. M. N. Srinivas, "Social Anthropology and Literary Sensibility," *Economic and Political Weekly* 33, no. 39 (Sept. 26-Oct. 2, 1998): 2528.
40. Ibid.
41. Ibid.
42. Stephen Davies, "Responding Emotionally to Fictions," *The Journal of Aesthetics and Art Criticism* 67, no. 3 (Summer, 2009): 270.
43. Ibid.
44. Ibid., 271.
45. Ibid.
46. Ibid.
47. Ibid.
48. F. Stuart Chapin III, "The Planet Is Shouting but Nobody Listens," in *Moral Ground: Ethical Action for A Planet in Peril*, ed. Kathleen Dean Moore and Michael P. Nelson (San Antonio: Trinity University Press, 2010), 77.
49. Barbara Kingsolver, *Flight Behavior*. (New York: HarperCollins, 2012), 122.
50. Ibid., 246.
51. Ibid., 247.
52. F. Stuart Chapin III, "The Planet Is Shouting but Nobody Listens," 77.
53. Robin Morris Collin, "Restoration and Redemption," in *Moral Ground: Ethical Action for a Planet in Peril*, eds. Kathleen Dean Moore and Michael P. Nelson, (San Antonio: Trinity University Press, 2010), 85-86.
54. William Gleason, "Looking Back to Look Ahead: Climate Change and US Literary History" in *Teaching Climate Change in the Humanities* (New York: Routledge, 2017), 209.

Chapter Two

The Taker-Leaver Paradigm

Cultural Representations in Contemporary Fiction

Contemporary authors who navigate cultural or personal identities as they relate to environmental and social justice can be a resource for defining better choices. As the chapter on Franzen's *Freedom* will explore more directly, fiction that confronts climate issues by developing characters' interactions with and responses to them provides a cultural guide for understanding how we should engage with a changing climate, including the consequences that accompany it. In any situation when a character has to choose how to respond, readers intuitively evaluate the efficacy of those choices. Educators have the responsibility to engage readers in texts that challenge them to think about how their own concepts of identity shape choices for interacting with the natural world. By extension, once readers learn to do this with fictional characters, applying a similar analytical process to real world decisions is a natural evolution. The extent to which individuals—from everyday citizens to responsible politicians to environmental scientists—act toward this goal will significantly affect our planet's future. These individual reflections—book by book, classroom by classroom—can be collectively transformative. In support of educators interested in this mission, secondary level fiction titles are presented here with the goal of organizing a hypothetical course unit that fosters reflection, as is expected of a literature study, but also action.

Depending on course structure, these texts could be handled individually as a class or, if time does not permit, then as literature circle options for reading groups. The stylistic content of these novels offers ample means for satisfying state and national standards in reading and literacy. Furthermore, titles have been chosen for their applicability to subject areas other than English—specifically history, science, psychology, and government, for example. A unifying practical focus will help readers concentrate on environmental or social commentary offered by the fiction. A collection such as the one proposed here could concentrate on, for example, a lens of witness literature, with objectives tailored to the concerns in each selection. This encourages students to analyze craft as well as content in determining authorial intent, while contributing to their own skills in these areas. Teachers might offer relevant contextual background, incorporate other disciplines, and encourage student research to extend collaborative learning. While this chapter offers only a sampling of contemporary fiction, many other texts illuminate similar issues—including titles commonly studied in American public education. Similarly, my analyses focus on a particularly relevant eco-concern of the novels, leaving much room for further study. Presenting two opposing cultural views of humanity's relationship to Earth, the stories discussed in this chapter illustrate an important premise for understanding the influence of culture on various ecological dilemmas that contribute to social inequities. Like the characters presented here, awareness of these opposing views urges readers to choose the perspective with which they wish to align themselves. With some guidance, we can hope they choose well.

Helpful in appreciating the crux of many environmental concerns, Daniel Quinn's *Ishmael* examines consequences of human exceptionalism. Proposed as the myth perpetuated by all "civilized" cultures, Quinn's story addresses why certain tribal groups living outside the confines of developed communities are the exception, exploring why exclusion from this adherence has been positive for them. In condemnation of this widespread cultural belief that human beings are above all other species, Quinn details some of the implications for our shared environment—such as destruction of land resources for agricultural endeavors, including competition with and decimation of other species. The focus of Quinn's novel—understanding what it means to be a Taker or a Leaver—forms the unifying lens through which I analyze each novel in this chapter. It is a helpful thematic thread that, once explained, can be applied to other fiction or nonfiction groupings that consider environmental concerns and human interactions with each other or other species.

Quinn's story begins with an unnamed narrator responding to what he initially perceives as a ridiculous ad in the paper: "Teacher seeks pupil. Must have an earnest desire to save the world. Apply in person."[1] Skeptical but intrigued, the narrator decides to "apply" to the ad if only to squash the concerned but defeated environmentalist left inside him. Instead of the human teacher he expects to find, he encounters a gorilla who had been named Ishmael during a previous captivity. Much to the narrator's surprise, Ishmael persuades him to stay and listen (after the narrator suspends his disbelief and submits to Ishmael's ability to communicate telepathically). As Quinn's narrator commits himself as pupil, so too does the reader vicariously commit.

Perhaps a nudge to readers who may have experienced similar despair, the narrator's intrigue overcomes his reluctance when he begins a quest for knowledge that will inspire again his forsaken youthful goal of saving the world.

Ishmael's tutoring sessions over time focus on deconstructing the myth of "Mother Culture" as a way of understanding the narrator's impression of living in "a civilizational system that more or less compels [us] to go on destroying the world in order to live."[2] As Ishmael explains, two narratives exist in humans' perception of their place in the world: the predominant one belonging to the "Takers" (civilized), the lesser known one to the "Leavers" (primitive).[3] As inquiry leads to the narrator's understanding of the Takers' story, Ishmael links its prevalence to human behavior toward the environment: "The mythology of your culture hums in your ears so constantly that no one pays the slightest bit attention to it. Of course man is conquering space and the atom and the deserts and the ocean and the elements. According to your mythology, this is what he was *born* to do."[4] The reference to Genesis here highlights the main contrast the novel strives to illustrate between cultures. Thomas Sancton frames it concisely in his essay "Planet of the Year: What on EARTH Are We Doing?":

> In many pagan societies, the earth was seen as a mother, a fertile giver of life. Nature—the soil, forest, sea—was endowed with divinity, and mortals were subordinate to it. The Judeo-Christian tradition introduced a radically different concept. The earth was the creation of a monotheistic God, who, after shaping it, ordered its inhabitants, in the words of Genesis: "Be fruitful and multiply, and replenish the earth and subdue it: and have dominion over the fish of the sea and over the fowl of the air and over every living thing that moveth upon the earth." The idea of dominion could be interpreted as an invitation to use nature as a convenience.[5]

As the narrator begins to understand this effect in myriad ways, he is able to apply it to current environmental dilemmas; so too might Quinn's reader.

As Ishmael explains, the perpetual reenactment of the Takers' myth—man must conquer his environment as a way of fulfilling his destiny— has landed humanity in an inevitably destructive role toward nature's resources and fellow species. As gorilla explains to man, the consequences of these actions—intentional or not—are a reality of the human experience. He frames the story humanity's actions tell as a systematic "taking" of resources, land, and life with no regard to social equity or ecological limits.

Later in their sessions, in a challenge to the narrator to define the laws that govern how we Takers live, Ishmael sardonically comments, "I have amazing news for you. Man is *not alone on this planet.* He is part of a community, upon which he depends absolutely."[6] As the narrator concedes, humans see themselves as exempt from the basic laws of behavior in this community of life because their "Mother Culture" determines that they are above it. Thus, the laws of nature that other species consistently follow regarding consumption of food, range of territory, social behaviors, are not recognized, much less honored, by people of Taker cultures.

The point Ishmael (and quite poignantly, Quinn) eventually makes is that these laws cannot be ignored without consequence. In exposing the myth of human exceptionalism, Ishmael demands that his pupil uncover the law of living that might save humanity from extinction. The clue to this discovery is that "only once in all history of this planet has any species tried to live in defiance of this law—and it wasn't an entire species, it was only one people, those I've named Takers."[7] As Ishmael explains, this group of Takers confronted and conquered Leaver cultures, spreading their defiance until the world became predominantly inhabited by them. After several days of thought, Quinn's narrator returns to Ishmael to report his findings; he has compared human behavior to that of various species' behavior in the wild, concluding that there are four behaviors characteristic of the Takers alone:

> This law that you have so admirably described defines the limits of competition in the community of life. You may compete to the full extent of your capabilities, but you may not hunt down your competitors or destroy their food or deny them access to food. In other words, you may compete, but you may not wage war.[8]

Ishmael further questions the narrator about the implications of species waging such wars, to which he concludes: "I'd have to say there would only be one form of life at each level of competition."[9] Thus, as Ishmael refers to it, this "peacekeeping" law protects diversity among and between species. The Takers' war on each other and other species is therefore disruptive to the whole system's order and balance. Additional implications of breaking this law—the legacy of the Takers—include land, air, and water pollution; species extinction; habitat destruction; resource depletion; and overpopulation. As Ishmael criticizes, this disorder has been created by the Takers through a "mythology of human superiority [that] justifies their doing whatever they please with the world."[10] That perceived rationale as justification, Quinn seems to assert, has premised the myriad ecological disasters contemporary culture faces.

In contrast, after a lengthy discussion about the treatment of man by gods in mythology, the development of agricultural societies' conflicts over land in biblical stories, and the sequence of evolution, Quinn's narrator comes to understand the main distinction between the Takers' story compared to the Leavers': "The premise of the Taker story is *the world belongs to Man*. . . . The premise of the Leaver story is *man belongs to the world*."[11] The metaphor Leavers follow, which is less self-centered than that of the Takers, is enacted, according to Ishmael, by such groups as "the Bushmen of Africa, the Alawa of Australia, the Kreen-Akrore of Brazil, and the Navajo of the United States."[12] Rather than viewing man as master of the world, the Leavers behave within a story that respects the law of community. The difference is simplified further in the namesake of each "Mother Culture" (as Ishmael calls it); Takers are guilty of consuming or collecting more than they need while Leavers will take only what they need, content to leave the rest. Ishmael primarily defines Takers throughout history as agriculturists, while Leavers are hunter-gatherers. In terms of contemporary solutions for current cultures, this distinction perplexes the narrator. Certainly it is not the impossible return to hunting-gathering that will save the world, he worries. Instead, Ishmael explains, it is a reversal of mind-sets in Taker culture that must be addressed. The challenge, articulated by the narrator, is how we convince the Takers that there are other ways to live that respect the natural environment as well as other species or competitive human groups.

In a most fitting directive, Quinn's Ishmael tells his student, who may as well be the reader, that "what you do is teach a hundred what I've taught you, and inspire each of them to teach a hundred."[13] Indeed, that is the approach that landed Quinn's novel into this Taker's hands. A proprietor of a small store that sells natural products and promotes nature's remedies for various ailments handed me the book while I was browsing for local honey. He cheerfully asked me to "read it and pass it on." I have done just that, believing, as does Robin Globus Veldman in "Narrating the Environmental Apocalypse: How Imagining the End Facilitates Moral Reasoning," that this small step can have serious traction. Her observations of participants in a "climate change-themed discussion group" include an interview of "'Sarah,' [who] had been interested in environmental issues for much of her life" but made significant changes to her lifestyle only after reading Quinn's work.[14] 'Sarah' explains the effect it had on her:

> . . . I bought about ten copies and started handing it out. . . . I've since read all his books. And so, upon reading that, I decided I wanted to learn how to grow my own food. . . and I just started 'Food Not Lawns'. . . I'm on a learning curve. . . .[15]

Veldman summarizes her reaction to reading *Ishmael*, stating that "it was a comprehension of the deep roots of the unfolding ecological crises and its potentially disastrous consequences that inspired her to make a series of substantial lifestyle changes."[16] Quinn's narrative can help educators, or any adult interested in this effort, examine these "deep roots" and thereby foster purposeful action—from passing the book on, to discussing environmentalism with friends and acquaintances, to planting food gardens or otherwise reducing Taker behaviors. Environmental apocalyptic narratives—both fiction and non—can turn individual awareness about environmental crises into concerned, effective community action.

Voices for justice can be heard from all parts of the world. Connecting these voices encourages a more complete assessment of the problems but also emphasizes the need for intercultural collaboration toward solutions. As defined by Quinn's gorilla tutor, characteristics of Taker versus Leaver cultures are also poignantly discernible in Luis Sepúlveda's *The Old Man Who Read Love Stories*. With an underlying theme of the power of stories, this conflict begins in an Ecuadorian jungle village terrorized by a grief-crazed female ocelot. A white tourist "hunting protected species out of season"[17] has killed ocelot cubs after wounding the father while their mother was out hunting for food. Upon returning to find her skinned cubs, the mother ocelot tracks and kills the hunter. When the ignorant mayor mistakenly accuses Shuar natives of murdering him, simply because they had found the body, the protagonist of the story—an old man named Antonio José Bolívar—cautions him:

> She scented and tasted human blood, and in her small animal brain we all murdered her litter, we all smell the same to her. Let the Shuar go. They must warn their people and the surrounding settlements. Every day that goes by will make the female more desperate and more dangerous, and she'll be after blood near the huts.[18]

The ocelot mother's natural grief—as well as readers' reaction to it— is compounded by Bolívar's criticism of the hunter, himself an unnatural presence in the Ecuadorian jungle: "That gringo was a real bastard! Look at the skins. Small and useless. Hunting on the eve of the rainy season, and with a shotgun! Look at the holes."[19] As Scott DeVries notes in "Swallowed: Political Ecology and Environmentalism in the Spanish American 'Novela de la Selva,'" a sense of environmentalism specific to jungle settings permeates the novel. Particularly in this opening scene, the opposition between cultures is clearly established: ". . . harmful activities include those of *gringo* hunters and *colono* gold seekers, whereas the Shuar community's knowledge of the land is held up as the preferred noncommercial model."[20] Further illustrating this oppositional paradigm, conflicts surrounding each culture's treatment of the natural world are developed through the protagonist's flashback reminiscences.

In the story that unfolds, readers learn how an attempt by settlers "promis[ed] large tracts of land and technical help" to "colonize part of Amazonia"[21] had been miserably thwarted by the harsh realities of life in the jungle. As in Ishmael's versions of "Mother Culture" myths, the old man reflects upon experiences which highlight an unnatural relationship with the earth. In his younger years, Antonio José Bolívar had been one of many disdainful agriculturalists trying to make a new life by taming the jungle with seeds and spade. The settlers (Takers) eventually have to be saved by the Shuar (Leavers) who "taught them to hunt, fish, build solid huts that would withstand the rains, and distinguish between edible and poisonous fruits. Above all, they taught them how to live in harmony with the jungle."[22] The eventual lesson for the settlers, who stubbornly ignored the natives, is that human-imposed order over the land is not sustainable:

> Despite what the natives had said, they sowed their first seeds, and soon found out for themselves that the soil was poor. The constant washing of the rains meant that the plants didn't receive the nourishment they needed and died without flowering, their growth stunted, or were devoured by insects.[23]

The effect on Antonio José Bolívar, who increasingly distances himself from new settlers and aligns his way of life with the Shuar natives, is a reversal of mind-set—from Taker to Leaver: "[He], who never thought about the word 'freedom' in the jungle, enjoyed limitless freedom. However much he tried to revive his old hatred [for the jungle], he couldn't help loving that world, and then the hatred faded as he was seduced by those vast expanses without frontier or owner."[24] This moving scene in the narrative nurtures thoughtful consideration of the meaning of existential freedom as it contrasts the economically-motivated freedoms we enact daily on the environment.

As John Terborgh concedes regarding such freedoms in *Requiem for Nature,*

> Whether we like it or not, tropical forests are worth more dead than alive. Nothing can save them short of a sea change in public opinion that registers not only in politicians' statements but also in their actions. Saving biodiversity will have to become a global obsession, not merely a pastime.[25]

Rather than loving "those vast expanses without frontier or owner," individuals or governments may sell them to the highest bidder without regard for future implications or perhaps, as in the case of many undeveloped nations, without an alternative to the financial windfall their destruction ensures. Addressing the personal "freedom" of land ownership, Terborgh proceeds with an argument that could enhance the study of novels like Sepúlveda's, complementing civics, ethics, or even math discussions:

> The U.S. tradition of absolute property rights will eventually be recognized for what it has already become, a detrimental anachronism, an impediment in the path of real progress toward sustainable development. The point can be made in the form of a simple ethical conundrum. The land belongs to the planet earth and is here forever. An individual lives for but a fleeting moment in time. Is it then right that an individual be allowed to degrade the land, making it unusable by future generations?[26]

He develops his point by contrasting an "economic system that espouses unending growth, discounts the future, and undervalues natural resources" with a "morality in tune with a truly sustainable society, [where] the idea that an individual could buy, own, and destroy for profit a forest of thousand-year-old trees would be completely abhorrent and unacceptable."[27] Studying historical and contemporary examples of actions that violate this precept in pursuit of economic gain could reinforce students' appreciation for the intricacies inherent in climate-related issues, specifically the interwoven injustices for both landscapes and people.

Reinforcing the possibility of shifting mind-sets, Sepúlveda's young settler grows into an awareness that centers on his respect for the land, accepting dependence on it rather than exercising ambition to conquer it. While he is never fully considered a native of the jungle, having once made a grave error in a revenge killing that ultimately prevents his acceptance by the Shuars, Antonio José Bolívar lives the remainder of his life as a Leaver. Through descriptions of his life in the jungle, Sepúlveda creates for readers a critique of Taker culture's behavior in previously unspoiled lands: "Both the settlers and the gold prospectors made all kinds of stupid mistakes in the jungle. They pillaged it unmercifully, so that some animals turned really hostile."[28] Marked by the human exceptionalism myth, this reference exposes the senseless abuse of resources, destruction of nature, and slaughter of animals characterized throughout history by Taker cultures' determination to conquer land. While the Shuars and the animals retreat further into the jungle, Antonio José Bolívar is left to watch helplessly while Takers "destroyed the jungle and constructed that masterpiece of civilized man: the desert."[29] The structure of the novel, by describing Bolívar's evolving perspectives, encourages reflection in that it "advocates an ethics of human-nature interaction after this '*ley shuar*' that rejects excessive use of resources and deforestation in favor of a lifestyle of conservation and sustainable consumption."[30] Like its protagonist, readers may be discouraged, even enraged, by the destructive Taker behavior which is so obviously juxtaposed to the more harmonious practices of the Shuar natives.

DeVries explains how the use of setting enhances these thematic concerns of the text, which offer multiple opportunities for extended analysis:

> This idealized vision of the *selva* as a nourishing and invigorating place complements the critique of development in the narrative, as many moments in Sepúlveda's novel describe several of the most central tenets of environmentalism: conservation, sustainability, critiques of deforestation, the idealization of nature, and the wisdom of indigenous groups.[31]

Educators might enhance the transformative experience of reading fiction by exploring appropriate accompanying nonfiction: for example, investigating the ethics of hunting among cultures; the ecological ramifications of forest and wildland destruction; the unique values of indigenous populations; the politics of imperialism, colonization, or industrialization. Concerns related to religion, economics, and sociology would also benefit the study, fostering intriguing interdisciplinary conversations.

Toward the conclusion of this short novel, Sepúlveda offers readers resolution to the mother ocelot's raging grief through an act of redemption by Antonio José Bolívar. He has been commissioned by the mayor to kill the ocelot for her predatory threat to humans in the villages. During the hunt, the ocelot unexpectedly leads the old man to her wounded mate. With apology, he puts him out of his misery: "Forgive me, friend. That lousy gringo did us all wrong."[32] Sensing her anguish, Antonio José Bolívar imagines the female, now out of sight, hiding in the trees and "shaken by sobs that were almost human."[33] The description acts as a compelling commentary of Taker culture's tendency to separate nonhuman animals from humans as a means of justifying callous behavior toward them. By acknowledging the ocelot's loss, Sepúlveda provides an opening to discuss philosophical perceptions of animal emotions as well as evolving scientific studies regarding them. From a cultural perspective, this might extend to analyzing different groups' hunting practices, approaches to species conservation, or beliefs regarding animals raised as pets versus those raised for food.

In a final act as hunter, the old man is forced to shoot the attacking female—who provokes her own death as if seeking solace for the deaths of first her offspring, now her mate. Perhaps as all Takers should, the old man "wept tears of shame, feeling unworthy, degraded, not at all the victor after the battle."[34] After abandoning the gun forever, Antonio José Bolívar returns to "his novels that spoke of love in such beautiful words they sometimes made him forget the barbarity of man."[35] The hope, of course, for instructors sharing this wonderful novel is that students will not so easily escape back into those oblivious Taker romances and delusions—that they too might fall in love with the jungle. By respecting his transformation, we might seek harmonious lifestyles in our own settings. By endeavoring to reach Antonio José Bolívar's level of contentment, we honor the type of freedom it symbolizes for this protagonist.

Another compelling advocate, Barbara Kingsolver introduces *The Poisonwood Bible*—a story of a southern Georgia Baptist family's mission trip to a village in the Belgian Congo—with Genesis verse 1:28, which speaks again to the premise of the Taker myth Ishmael highlights:

> *And God said unto them, / Be fruitful, and multiply, and replenish the earth, / and subdue it: and have dominion / over the fish in the sea, and the fowl of the air, / and over every living thing that moveth upon the earth.*

Narrating the first chapter of a generational story, Orleanna Price alludes to this biblical instruction as part of a powerful reflection on the history of Africa and America's experience in it:

> We aimed for no more than to have dominion over every creature on the earth. And so it came to pass that we stepped down there on a place we believed unformed, where only darkness moved on the face of the waters. Now you laugh, day and night, while you gnaw at my bones. But what else could we have thought? Only that it began and ended with us. What do we know, even now? Ask the children. Look at what they grew up to be. We can only speak of the things we carried with us, and the things we took away.[36]

Kingsolver's chilling development of this reference to Genesis speaks to Quinn's description of human behaviors powered by a belief in the destiny of man to conquer the earth—a belief which we have no logical right to hide behind. Just as Tim O'Brien does of the Vietnam War, Kingsolver's criticism of the precept is clear throughout the text as female members of the Price family tell their versions (in contrast to Nathan's) of life spent in Africa as Takers among a Leaver culture.

Weaving political, social, and environmental history through chapters with biblical names, Kingsolver's narrators indict various levels of complicity—familial, religious, cultural—within Taker culture. Jeannette E. Riley explores this critique in "The Eco-Narrative and the Enthymeme: Form and Engagement in Environmental Writing":

> Orleanna's comments challenge readers to conceptualize a different story and truth—to complete Kingsolver's argument—that the African people and the land are being (and have been) oppressed by the colonizer, represented by the Prices and by their church and government.[37]

Engagement with this fictional family's conflict engenders empathy for a Leaver culture and supports readers' interaction with actual truths that Kingsolver confronts about external influences on Africa throughout history into the present. How audiences contemplate knowledge regarding those truths defines the transformative nature of texts like this. For example, the way readers individually respond to characters' conflicts in the narrative may drive post-reading research and conversation that hopefully leads to action. As Riley notes about stylistic choice, such rhetorical strategy in environmental literature "[calls] [readers] to act on behalf of or to integrate the teachings of the environment into their lives."[38] This is especially encouraged by the changes some of the characters make in their own lifestyles once they become cognizant of their unwitting participation as Takers. Kingsolver's narrative works particularly well in its presentation of how interaction within the family unit directly impacts interaction with others in particular environments.

From Reverend Price's adamant manipulation of the land to plant a garden (which never produces) to his fervent attempts to Christianize the African village by baptizing all its children, readers witness a misguided urgency to conquer land and people in the name of God. As Riley observes, "the Price family's interactions with the land itself . . . also enthymemetically connect patriarchal power and domination over nature to a disregard for the cultural practices of the bioregion."[39] We can see, for example, Nathan Price perpetuating religious perceptions through his ignorant persistence to dominate the soil of Africa by creating a garden.[40] His violent behavior toward the land he attempts to cultivate and his arrogant dismissal of African native Mama Tataba's advice are witnessed by his daughter Leah, who later will understand how this cultural ignorance is a flaw of her father's belief system. For readers, his behavior is consistent with the domineering behavior exhibited toward his family as well as the African villagers he intends to save:

> It does not occur to Nathan that the environment in the Congo might differ from that of the southern United States; nor does it occur to Nathan that a villager, one born and raised on the land, might know the land better than he does. These assumptions mark him as "colonizer" and "oppressor" to readers who refuse to condone Nathan's cultural arrogance and desire to control his environment.[41]

Since these labels also characterize his role as father, readers are reminded of how intricately woven such interactions are—from self, to family, others, and environments. Fittingly, "[Nathan's] demonstration garden will bear no fruit as this garden, an allusion to the garden of Eden, fails due to his pride and inability to comprehend the new land he attempts to dominate."[42] As his garden fails, so too do efforts with his family and affiliates (the African villagers), who grow to disdain him. As readers, we see how pervasively this Taker mind-set tries to justify itself as sanctioned by God Himself. The symbolic garden scenes offer plentiful opportunities to reference similar fictional econarratives, as well as any number of historical, political, or anthropological texts about human interactions with new lands and people (such as Adam Hochschild's *King Leopold's Ghost*, Alan Weisman's *The World Without Us*, Jared Diamond's *Collapse: How Societies Choose to Fail or Succeed*).

Another aspect of the text, which helps to explain Nathan Price's attitude, is offered by a particularly intriguing character—one of the Price daughters, Adah, who has been disabled from birth. Through her, Kingsolver presents a child's perspective on the United States' political attitude toward Africa during the Eisenhower administration, specifically its plan to kill elected African president Lumumba: "I crept to my bed and wrote what I had seen and heard, then wrote the ending backward. Stared at the words in my notebook, my captive poem: Redrum sekil oh weki ekil ew."[43] Likening her conclusion—we like ike who likes murder—to her confusion with the missionary presence her father leads, Adah decides it is not surprising: "How is it different from Grandfather God sending the African children to hell for being born too far from a Baptist Church? I should like to stand up in Sunday school now and ask: May Africa talk back? Might those pagan babies send *us* to hell for living too far from a jungle?"[44] Adah's perspective is one of innocent objectivity that allows her to perceive what is rather than what is believed by cultural myth. Reality is cemented by a crucial moment in the story when the African village the Price family resides in is overtaken by ants that consume everything in their path. Left behind by a mother forced to choose one child to carry to safety, disabled Adah finds herself crushed under "a river of people saving themselves." She was surprised to find that "even the crooked girl believed her own life was precious" and worth saving. Her reflection on this moment is perceptive: "That is what it means to be a beast in the kingdom."[45] Readers may find this passage alarming as it likens human survival to that of all living "beast[s]," thus challenging our perceptions about how lives—human and nonhuman—tend to be hierarchically evaluated in modern culture.

A climate-related connection to Adah's realization might include consideration of how developed nations are valuing or devaluing people and landscapes currently being impacted by severe weather and rising sea levels. How have and will wealthier nations treat those displaced from their homes? When the choice is forced, who decides what lands will be protected, which people will receive resources? Documentaries like Michael Nash's *Climate Refugees* and the American television series *Years of Living Dangerously* help illustrate the human story behind weather-related facts we might study in science, or resources versus population statistics in math. Mature readers might consult meticulously researched texts like Mark Lynas's *Six Degrees: Our Future on a Hotter Planet* or Elizabeth Kolbert's *The Sixth Extinction: An Unnatural History*, which rely on climate scientists' data to report on historical, current, and future consequences of climate change.

Adah's perplexing awareness of her own identity as a survivor haunts her when she later becomes a doctor in the United States: "Who was I, vowing calmly among all these necktied young men to steal life out of nature's jaws, every old time we got half a chance and a paycheck?"[46] Her momentous understanding of what it means to be "a beast in the kingdom," driven by a will to live at all costs, also acknowledges the environmental influences and implications of that instinct. This exasperating juxtaposition of what survival and life means in an environment like America—where "we sit among such piles of leftover protein we press it into cakes for the pets" and "pay soothsayers and acrobats to help lose our weight"[47] —against the same effort in Africa becomes ultimately irreconcilable for Adah:

> Overpopulation has deforested three-quarters of Africa, yielding drought, famine, and the probable extinction of all animals most beloved by children and zoos. The competition for resources intensifies, and burgeoning tribes itch to kill each other. For every life saved by vaccination or food relief, one is lost to starvation or war. Poor Africa. No other continent has endured such an unspeakably bizarre combination of foreign thievery and foreign goodwill. Out of sympathy for the Devil and Africa, I left the healing profession.[48]

Adah's experience as an American child living in an African village offers privileged readers an opportunity to see what is rather than what our cultures have passed along as acceptable paradigms. Adah's realization reminds readers of Ishmael's caution against personal excesses that make us culpable for what has become of the planet.

Kingsolver does not overlook the irony that Taker culture has perpetuated the human myth by, in fact, creating the Africa Adah describes. As in Quinn's and Sepúlveda's novels, Kingsolver traces the myth back to the conflicting metaphorical identities of Taker and Leaver cultures. In a chapter that Leah Price narrates, "Song of the Three Children," Kingsolver explores the history of a world's injustice to Africa through a Taker culture's merciless conquering of a land where Leavers once prospered. Again, identity and purpose are tied to the land and the myths its people foist upon it:

> If only a river could go uncrossed, and whatever lay on the other side could live as it pleased, unwitnessed and unchanged. But it didn't happen that way. The Portuguese peered through the trees and saw that the well-dressed, articulate Kongo did not buy or sell or transport their crops, but merely lived in a place and ate what they had, like the beasts of the forest. In spite of poetry and beautiful clothes, such people were surely not fully human—were *primitive*; that's a word the Portuguese must have used, to salve their conscience for what was to come. Soon the priests were holding mass baptisms on shore and marching their converts onto ships bound for sugar plantations in Brazil, slaves to the higher god of commodity agriculture.[49]

As the gorilla points out for his human student in *Ishmael,* the difference between these groups of people was one of ethos: one group took from the land what was needed when it was needed; the other exacted dominance over the land to produce in excess what it wanted. Leah's repudiation of ethnocentric colonialism's lasting impact mirrors Ishmael's condemnation of agriculturists' invasion of hunter-gatherer cultures. Much more than an acknowledgment of right or an indictment of wrong, the distinctions in each author's story point to a philosophical opposition of lifestyles. Specifically, each author demonstrates that the appearance of Taker culture in a land initiates its immediate, violent shift toward the Taker myth; the invariable result is a damaged environment and human psyche. As Adah and Leah come to understand, one's stance on which lifestyle becomes our cultural myth is determined by one's location as much as experience.

Significantly, it is Leah who stays in Africa. Her sister Rachel also stays but escapes with her husband of convenience to South Africa, where she continues to enact the Taker myth from which she came. That Leah stays is important because as a child eager to please and gain his affection, she had been most entangled with her father's missionary goals. In contrast to what Nathan Price might expect from his most dutiful of daughters, Leah forsakes her religion, her father, and her homeland choosing instead to marry Anatole—a teacher in the village who used to translate Price's sermons. Like the old man in Sepúlveda's novel, Leah now becomes part of the opposing myth. The life she chooses is not an easy one—"here you know what a seed is for, or you starve"[50] —but she finds shameful contempt for her homeland of excess. Leah seeks redemptive solace in living off African land while teaching others how to do the same. A message gleaned from her recognition is that most of us, like Nathan, never reach it:

> . . . it would be the simple human relief of knowing you've done wrong, and living through it. Poor Father, who was just one of a million men who never did catch on. He stamped me with a belief in justice, then drenched me in culpability, and I wouldn't wish such torment even on a mosquito.[51]

If empathic readers have not seen themselves in Adah's and Leah's transformations to Leaver culture or in Rachel's and Nathan's continued roles as Takers, perhaps Orleanna's story rings true. In constructing the family narrative, this mother faces painful culpability in her marriage, religion, and patriotism. Illustrating the interconnections of our various identities, Elaine R. Ognibene explores these relationships in "The Missionary Position: Barbara Kingsolver's *The Poisonwood Bible*." Confronting the power of words themselves, Orleanna's "Exodus" chapter exemplifies the influence of narrative:

> She begins by defining the need to understand the deceptive nature of words, a recurring theme in the novel: '*Independence* is a complex word in a foreign tongue. To resist occupation, whether you're a nation or merely a woman, you must understand the language of your enemy. *Conquest* and *liberation* and *democracy* and *divorce* are words that mean squat . . . when you have hungry children . . .' (Kingsolver 1999, 383). Orleanna's wisdom about the space between words moves her to change. She accepts responsibility for her complicity and acquires the words for her story. For Orleanna, telling her story is a syncretic process, as she aims to reconcile what has gone before.[52]

Orleanna's reflections as citizen, wife, mother are layered in the complex cultural and political contexts Kingsolver explores throughout the text. While never justifying Orleanna's participation in the Taker myth, Kingsolver here acknowledges that options are not always as simple as distinguishing right from wrong.

In the case of environmental and social justice issues, these complexities may prevent basic awareness—much less meaningful action. By confronting the nature of culpability—through, in this case, discussions in history and literature courses—teachers could invite students to examine their role in situations the Price women critique and encourage simple changes in everyday practices. A generation that one by one becomes cognizant of how personal choice can impact change—from engaging in local concerns, to voting in elections, to participating in large-scale reform—has powerful potential. Advanced students, with confident teachers, will take this cognizance a step further by evaluating more complex culpabilities addressed in the novel. In *Occupy Education*, sustainability educator Tina Lynn Evans summarizes her rationale for engaging with students about how these layered frameworks induce enforced dependency, a key tenet to the critical pedagogy of sustainability education she advocates. Though specifically addressing America, her assertions are particularly relevant for an ecocritical study of Kingsolver's text and numerous others in which thematic concerns focus on juxtapositions between developed and undeveloped nations:

> Given the breadth and depth of U.S. complicity in perpetrating the sustainability crisis and the crisis of capitalism itself, I propose that U.S. citizens should engage in a form of self-reflexivity that critique their collective agency as a nation. We in the U.S. have contributed in many ways—willingly and knowingly or not—to widespread collective violence and ecocide ranging from war making to consumerism, and our nation has been the prime mover in creating systems of enforced dependency worldwide. The critical pedagogy of sustainability must engage students in critique of the U.S. as an empire in order to elucidate the mechanisms of enforced dependency in the world-system. Students should be encouraged to question the sustainability of this dependency and to envision and begin to create alternatives.[53]

Importantly, these reflections should be steered carefully to avoid shaming or offending students. The goal is not to place blame but to recognize the hegemonic mind-sets which inform cultural behaviors we find morally offensive, particularly those we unwittingly support when we shop, invest, or vote. There is little hope of changing such a pervasive force if education (which, after all, helped create it) does not safely engage students in confronting it.

Another similar indictment of Taker culpability is witnessed through a young Native American's return to the United States after World War II in Leslie Marmon Silko's *Ceremony*. Veteran and escaped hostage of the Japanese, Tayo struggles to resume a normal life on the Laguna Pueblo reservation. In his attempts to find solace, Tayo reveals the Taker-Leaver mythologies that have been and are being enacted in America. His trauma is characterized by memories of abandonment by his mother, by rejection from predominant culture, by violence and captivity in war, and by revelations of a colonized history. As Tayo seeks refuge in his cultural heritage through preserving traditions, a central trauma is uncovered as the root of his (as well as his cultural peers') alienation from society, particularly from the conventionally normal behaviors in which he is encouraged to participate. It begins in a reflection Tayo has as he looks over the drought ravaged land of his reservation:

> 'You see,' Josiah had said, with the sound of the water trickling out of the hose into the empty wooden barrel, 'there are some things worth more than money.' He pointed his chin at the springs and around at the narrow canyon. 'This is where we come from, see. This sand, this stone, these trees, the vines, all the wildflowers. This earth keeps us going.'[54]

Josiah goes on to explain to Tayo that it is not the elements that should be cursed for the drought but rather the people who are to blame for forgetting how to live properly on the land.[55] Throughout the novel, Tayo cannot identify in his memories the true enemies of a war he did not understand, nor is he able to connect with Native American friends who seek relief in alcohol, violence and the rejection of Indian traditions. In his self-imposed solitude, readers learn the poetic stories of Tayo's ancestors which eventually help him identify the source of his angst. His postwar identity crisis parallels Josiah's claim that the drought is a result of people forgetting their relationship with the land, a negative shift in mind-set attributed to the "Leavers" (Native American) versus "Takers" (European) conflict in America. Further emphasizing this dichotomy, a reminiscence of Josiah's hunting stories—ones that tell of a balance and respect for land, animals, and the sacrifice of death—leaves Tayo confronting the environmental realities of a violently colonized past and its implications.

Describing land once inhabited by his "Leaver" ancestors, Tayo laments its early 1900s "settlement" by white culture "Takers":

> . . . the loggers had come, and they stripped the canyons below the rim and cut great clearings on the plateau slopes. The logging companies hired full-time hunters who fed entire logging camps, taking ten or fifteen deer each week and fifty wild turkeys in one month. The loggers shot the bears and mountain lions for sport. And it was then the Laguna people understood that the land had been taken, because they couldn't stop these white people from coming to destroy the animals and the land.[56]

Significantly, Tayo's recognition of this cultural history prompts a symbolic defiance of it. His memory is of lands once owned by Native Americans, specifically those ancestors of the reservation Tayo inhabits. At one point he is alone on horseback searching for Josiah's cattle that disappeared while grazing, presumably wandering off reservation land into those desecrated mountain areas. Importantly, Tayo's closeness with his mare and the land they traverse is juxtaposed by the cruel imagery of the fence boundary he ultimately finds the cattle behind:

> The white man, Floyd Lee, called it a wolf-proof fence; but he had poisoned and shot all the wolves in the hills, and the people knew what the fence was for: a thousand dollars a mile to keep Indians and Mexicans out; a thousand dollars a mile to lock the mountain in steel wire, to make the land his.[57]

Despite this understanding of the unnatural boundary separating him from the cattle he seeks (and rightfully owns), Tayo is hesitant to assume the cattle are stolen because he finds them on a white man's land. In an ironic, frustrated recognition of his own hypocrisy, which is full absorption of the prejudice he despises—acknowledging that he would not hesitate to assume an Indian or Mexican stole the cattle—Tayo approaches the symbol of hatred with fury and resolve: "The lie. He cut into the wire as if cutting away at the lie inside himself . . . as long as people believed the lies, they would never be able to see what had been done to them or what they were doing to each other."[58] The complexity of Tayo's emotional awareness in this scene suggests that his postwar experience of isolation and his present engagement with deep loss are intricately tied to cultural histories he is beginning to understand. Sharon Holm, in "The 'Lie' of the Land: Native Sovereignty, Indian Literary Nationalism, and Early Indigenism in Leslie Marmon Silko's *Ceremony*," explores the politics of this "lie":

> The complicity of the Indians in the "loss" of their land—what the text repeatedly identifies as the hegemonic "lie" underpinning the dominant power structure—and that "lie's" foundational function at the heart of a reputedly inclusive American nationalism, *e pluribus unum*, is, at this stage, revealed to be bound up with a guilt-ridden negotiation of Native identity. Here *Ceremony* accurately reflects the divided and confused loyalties resulting not only from the hangover of a powerful World War II American patriotism but also from various postwar policies such as termination and relocation, policies that, as Chadwick Allen points out, were designed to turn "indigenous 'communities of descent' based on kinship, shared culture, and connection to specific areas of land into Western-style 'communities of assent' based on individual choice of association."[59]

As with Kingsolver's Leah, Tayo's recognition of this context is transformative. In his resolve, he begins to connect layers of alienation and diminishment with the more personal indoctrination of education. His reflection on the education of Indians in a predominantly white system illuminates the more subtle undermining of Leaver culture: "Like the first time in science class, when the teacher brought in a tubful of dead frogs, bloated with formaldehyde, and the Navajos all left the room; the teacher said those old beliefs were stupid."[60] Later in the text, Tayo directly connects the "lie" he has recognized to a "destroyer" culture, realizing that "the destroyers had tricked the white people as completely as they had fooled the Indians . . ."[61] As Tayo perceives it, the lie has become culture. Leavers have become Takers but the pervasive nature of cultivating such a manipulative shift causes all to forget the lie's origin.

Silko illustrates manifestations of this transformation in the final scenes of the text when Tayo visits Trinity site—reservation land that had been purchased and destroyed for the purpose of testing the atom bomb. Reflections on World War II include Tayo's inability to tell enemy from friend both abroad and at home; Silko connects this with the devastation of land by nuclear poisoning:

> From that time on, human beings were one clan again, united by the fate the destroyers had planned for all of them, for all living things; united by a circle of death that devoured people in cities twelve thousand miles away, victims who had never known these mesas, who had never seen the delicate colors of the rocks which boiled up their slaughter.[62]

There could be no distinction for the young Tayo among Japanese, American, or Native American in a war that positioned them all as destroyer. His alienation and subsequent inability to cope within postwar culture is thus explained:

> He cried the relief he felt at finally seeing the pattern, the way all the stories fit together—the old stories, the war stories, their stories—to become the story that was still being told. He was not crazy; he had never been crazy. He had only seen and heard the world as it always was: no boundaries, only transitions through all distances and time.[63]

Silko's point in this epiphany is that entrenchment in Taker culture cannot be absolved by accusations of blame but rather demands a shared cultural shift to a more ethical mind-set. The task, as we have seen, is to transform awareness from reflection to action. High school teacher Nancy Gilderhus suggests such an experience with two of her students' revelations after studying *Ceremony*:

> Aaron: This novel made me think about and question our Western traditions. This not only made me realize our wrongdoing but to detest it.

> Mary Beth: This is the most spiritual book I have ever read. What an ending when Tayo refuses to fight Emo [Tayo's foil]. I know Tayo will be at peace. That's comforting.[64]

While we are not always privy to the ways our students subsequently act (or not) on learning experiences like these, their engagement suggests the possibility for change. Gilderhus's own reflection on her intention for the novel offers powerful reason to include it in any humanities course: "Barbara Kingsolver best states my hope for teaching *Ceremony* as we approach the twenty-first-century: 'The possibility that kids might one day grow up to be neither the destroyers nor the destroyed' (Kingsolver 1990, 299)."[65] With a global audience at their technological fingertips, all readers of twenty-first-century generations are in a position to make this reality.

Yet another set of Takers appears in Don DeLillo's *White Noise.* Jack Gladney is a history professor in a small town whose family experiences the stereotypical onslaught of overstimulation created by a Taker world of conveniences, leisure, and technology. Characters are constantly subject to and contributors toward their culture's intrusive "white noise" which threatens to prevent any true form of intimacy. Readers are invited to examine the insecurities of the protagonist's daily life with his fourth wife Babette, his colleagues, and his four children by different complex women. His marriage seems held together mainly by their shared, all-consuming fear of dying. The two imagine death lurking around every corner, only fueling their anxiety by contemplating which of them will die first and leave the other alone. In a bizarre chapter called "The Airborne Toxic Event," the family experiences an evacuation due to an unidentified black cloud moving into the town. Ironically, Jack and Babette remain relatively calm as they load their children into the family station wagon. Heading out of town amongst a horde of neighbors attempting the same, Jack reflects on their appraisal of the threat:

> It is surely possible to be awed by the thing that threatens your life, to see it as a cosmic force, so much larger than yourself, more powerful, created by elemental and willful rhythms. This was a death made in the laboratory, defined and measurable, but we thought of it at the time in a simple and primitive way, as some seasonal perversity of the earth like a flood or a tornado, something not subject to control. Our helplessness did not seem compatible with the idea of a man-made event.[66]

The "man-made event" reminds us of Silko's critique of a destroyer culture that willfully acts with disregard toward anything that stands in its way. In this case, the motivation relates to the elimination of insect pests rather than human enemies. During the emergency aid care provided post-evacuation, Jack learns he has been exposed momentarily to the toxic chemical Nyodene

D. which is explained by his son as "a whole bunch of things thrown together that are byproducts of the manufacture of insecticide. The original stuff kills roaches, the byproducts kill everything leftover. A little joke our teacher made."[67] Jack's fear of death is now driven by the unknown effects of his exposure, which he keeps from the family. The reader's fear should lie in DeLillo's critique of the "noise" produced by humanity's tunnel-vision quest for convenience, just as descriptions of the airborne toxic event should recall Rachel Carson's *Silent Spring*.

Tellingly, as the town and Jack's family moves forward from the toxic event, Jack is disturbed yet strangely comforted by the normality soon resumed in their town, aside from the now mandatory evacuation simulations in which a few of their children participate. As he notes in comparison to the disrepair of some of the town's homes, "the supermarket did not change, except for the better. It was well-stocked, musical and bright. This was the key, it seemed to us. Everything was fine, would continue to be fine, would eventually get even better as long as the supermarket did not slip."[68] The commentary here is likened to the distracting, even numbing effect of the white noise stimulation that bombards our everyday lives. Concern over a toxic event or spill like the one experienced by Jack's town is mitigated by the normalcy with which those around us move forward and, inevitably, forget. We turn up the volume, change the station, go to the supermarket. As Jack notes, the reassurance of surplus housed by brightly lit, well-organized grocery stores is a security that assuages our fear and prompts our return to the white noise.

This sheltering noise itself is complicit in our forgetting. Babette comments in response to a neighbor's heightened sense of dread regarding this airborne toxic event and others like it:

> Every day on the news there's another toxic spill. Cancerous solvents from storage tanks, arsenic from smokestacks, radioactive water from power plants. How serious can it be if it happens all the time? Isn't the definition of a serious event based on the fact that it's not an everyday occurrence?[69]

Babette's remark echoes the white noise of current Taker culture. We are experiencing an increasing number of disasters—both man-made and natural—but our concerned response to each, like Babette's, lasts only to the next "event." Or, we immerse ourselves in the stimuli of our lives desperately seeking the experience of Jack's youngest son Wilder in the supermarket: "the world was a series of fleeting gratifications. He took what he could, then immediately forgot it in the rush of a subsequent pleasure. It was this forgetfulness [Jack] envied and admired."[70] It may be natural to envy it in one so immature, yet when adults old enough to know better do so our response is one of condemnation. Readers are later disgusted when Jack himself seeks fleeting pleasure in a fast-food experience, a shopping spree, and ultimately an obsessive quest to access a drug purported to reduce one's fear of death. English professor Cornel Bonca, in "Don DeLillo's White Noise: The Natural Language of the Species," expresses the effect of teaching this novel:

> . . . the novel seems to draw out a certain buried awareness in my students that the most familiar aspects of their lives—shopping malls, television, families, and the languages of these things—harbor deep and resonant mysteries. It affects them, I think, as a sustained defamiliarization of their own lives. After reading it, it is (or should be) impossible to shop in a supermarket the same way, to watch a televised disaster the same way, even—and this is crucial—to listen to a baby's cry the same way. Preposterously funny, immediately accessible, yet deeply sophisticated on a formal and stylistic level *White Noise* is one of the few novels capable of mastering—perhaps taming—our schizoid confusions about the mass media experience. It is a novel which, because of its wide-ranging explanatory power and uncanny compassion, some *helps.*[71]

In terms of identity, this American novel wreaks havoc on the American dream—there is no questing for happiness or seeking fulfillment, only a superficial attempt to stop fearing man-made death so that characters can live lives of profligate consumption in "the white noise of advertising and mass media."[72]

Naomi Klein's assessment of lack of media coverage on the man-made death that is climate change similarly establishes this culprit: "Why do liberal media outlets still segregate stories about melting ice sheets in their "green" sections—next to viral videos of cuddly animals making unlikely friendships?"[73] In terms of coverage, she notes a disturbing decline that should also earn our outrage: " . . . in 2007, the three major U.S. networks—CBS, NBC, and ABC—ran 147 stories on climate change; in 2011 the networks ran just fourteen stories on the subject."[74] This again speaks to the importance of how and if the public attains information about climate change. If shifting mind-sets is a strategy we should embrace in fighting long-term consequences of climate change, which have been convincingly, consistently validated by 99 percent of scientists around the world, education has to be successful where the media have failed. Otherwise, this "white noise" will be our downfall. As Bonca suggests, DeLillo's novel is not merely critical but more importantly "realizes that it is in that noise that our terrors and longings can be read."[75] This too is part of the human narrative that we can deconstruct with students because, like Jack, fear of death or the unknown drives our behaviors. In the case of climate change—an overwhelmingly consistent threat like the airborne toxic events alluded to in the novel—responses range from despair to denial to panic. How we come to terms with our own white noise is part of what it means to be human, but in the face of this particular threat time will not forgive.

Fostering empathic reaction to text portrayals of environmental dilemmas that mirror real world situations is a place to start the cultural mind shift needed to confront climate change realities. As contemporary authors provide material through which we might better understand our position regarding Taker-Leaver cultures, it is incumbent upon educators to facilitate meaningful interactions that encourage reflective discussion and debate about them. The protagonists of the novels mentioned throughout this chapter are no less important than the characters most often studied in literature courses around the world. An important difference, however, is that characters like Ishmael's student, Leah Price, Antonio José Bolívar, Tayo, and Jack Gladney are like today's audience in their struggle to understand identity as it relates to contemporary, changing environments and threats. Equally, there are those who identify instead with the associated Taker characters in each of the texts—Nathan Price, or Sepúlveda's "gringo," for example. An educator's responsibility is to both groups. Indeed, while most of us may never be faced with decisions of such import as the literary figures we analyze, all of us are faced with environmental Taker versus Leaver decisions multiple times a day. How appropriately we maneuver through these daily choices is primarily determined by awareness of their complexities. Armed with a Taker-Leaver paradigm in which to frame our decisions and their ramifications, this engagement with literature could help students transform destroyer culture to live in a way that respects the natural environment and all its inhabitants.

After all, as Jared Diamond asserts in his introduction to *Collapse: How Societies Choose to Fail or Succeed*, we have reached a turning point where information demands response:

> I don't know of any case in which a society's collapse can be attributed solely to environmental damage: there are always other contributing factors. . . . Four of those sets of factors—environmental damage, climate change, hostile neighbors, and friendly trade partners—may or may not prove significant for a particular society. The fifth set of factors—the society's responses to its environmental problems—always proves significant.[76]

Schools can play a meaningful, lasting role in influencing this fifth set of factors by educating students who will choose to succeed.

NOTES

1. Daniel Quinn, *Ishmael* (New York: Bantam Books, 1995), 4.
2. Ibid., 25.
3. Ibid., 39.
4. Ibid., 74.
5. Thomas Sancton, "Planet of the Year: What on EARTH Are We Doing?," *TIME*, (January 2, 1989).
6. Daniel Quinn, *Ishmael,* 99.

7. Ibid.
8. Ibid., 129.
9. Ibid.
10. Ibid., 146.
11. Ibid., 239.
12. Ibid., 147.
13. Ibid.
14. Robin Globus Veldman, "Narrating the Environmental Apocalypse: How Imagining the End Facilitates Moral Reasoning Among Environmental Activists," *Ethics and the Environment*, 17, no. 1 (Spring 2012), 14.
15. Ibid.
16. Ibid., 15.
17. Luis Sepúlveda, *The Very Old Man Who Read Love Stories* (New York: Harcourt Brace, 1995), 20.
18. Ibid., 19.
19. Ibid.
20. Scott DeVries, "Swallowed: Political Ecology and Environmentalism in the Spanish American 'Novela de la Selva,'" *Hispania*, 93, no. 4 (December 2010), 537.
21. Luis Sepúlveda, *The Very Old Man Who Read Love Stories*, 30.
22. Ibid., 3 –3.
23. Ibid., 33.
24. Ibid., 34.
25. John Terborgh, *Requiem for Nature* (Washington, D.C., Island Press, 1999), 140.
26. Ibid., 155.
27. Ibid., 156.
28. Luis Sepúlveda, *The Very Old Man Who Read Love Stories*, 49.
29. Ibid.
30. Scott DeVries, "Swallowed," 536.
31. Ibid., 536.
32. Luis Sepúlveda, *The Very Old Man Who Read Love Stories*, 124.
33. Ibid., 125.
34. Ibid., 130.
35. Ibid., 131.
36. Barbara Kingsolver, *The Poisonwood Bible* (New York: Harper, 1998), 10.
37. Jeannette E. Riley, "The Eco-Narrative and the Enthymeme: Form and Engagement in Environmental Writing," *Interdisciplinary Literary Studies*, 10, no. 2, Ecopoetics and the Eco-Narrative (Spring 2009), 86.
38. Ibid., 82.
39. Ibid., 86.
40. Ibid.
41. Ibid., 87.
42. Ibid., 87.
43. Barbara Kingsolver, 238.
44. Ibid.
45. Ibid., 306.
46. Ibid., 527.
47. Ibid.
48. Ibid., 528.
49. Ibid., 522.
50. Ibid., 524.
51. Ibid., 525.
52. Elaine R. Ognibene, "The Missionary Position: Barbara Kingsolver's *The Poisonwood Bible*," *College Literature*, 30, no. 3 (Summer, 2003), 23.
53. Tina Lynn Evans, *Occupy Education: Living and Learning Sustainability* (New York: Peter Lang Publishing, 2012), 226.
54. Leslie Marmon Silko, *Ceremony* (New York: Penguin Books, 1977), 45.

55. Ibid.
56. Ibid., 186.
57. Ibid., 188.
58. Ibid., 191.
59. Sharon Holm, "The "Lie" of the Land: Native Sovereignty, Indian Literary Nationalism, and Early Indigenism in Leslie Marmon Silko's "Ceremony," *American Indian Quarterly*, 32, no. 3 (Summer, 2008), 259-260.
60. Leslie Marmon Silko, *Ceremony*, 194.
61. Ibid., 204.
62. Ibid., 246.
63. Ibid.
64. Gilderhus, Nancy, "The Art of Storytelling in Leslie Silko's 'Ceremony,'" *The English Journal* 83, no. 2 (Feb. 1994), 72.
65. Ibid.
66. Don DeLillo, *White Noise* (New York: Viking Penguin Inc., 1985), 128.
67. Ibid., 131.
68. Ibid., 170.
69. Ibid., 174.
70. Ibid., 170.
71. Cornel Bonca, "Don DeLillo's White Noise: The Natural Language of the Species," *College Literature* 23, no. 2 (June 1996), 25–26.
72. Ibid., 40.
73. Naomi Klein, *This Changes Everything: Capitalism vs. The Climate* (New York: Simon & Schuster, 2014), 62.
74. Ibid., 34.
75. Cornel Bonca, "Don DeLillo's White Noise," 40.
76. Jared Diamond, *Collapse: How Societies Choose to Fail or Succeed* (New York: Penguin Books, 2005), 11.

Chapter Three

Popular Science Fiction and Fantasy

Fostering International Perspectives

The choice to succeed or fail lies in humanity's ability to cast aside a Taker's mind-set in favor of a Leaver's. Literature that illustrates these opposing models can guide important engagement with the politics of a changing world by prompting contemplation of cultural mind-sets we choose to accept. Stories also remind readers that the way humans interact with physical environments is influenced by the quality of their personal relationships. In an increasingly mobile, technological, global society, these considerations of morality and connection become ever more significant. With access to education improving worldwide, one of humanity's best weapons for ameliorating numerous dilemmas related to the clash of Taker-Leaver mind-sets is climate fiction. As we have seen, the art of storytelling can prompt empathy for a culture's particular concerns—that which their artists witness—but can also move readers to appreciate those of cultures different than their own. This chapter explores how interdisciplinary studies of popular science fiction, speculative fiction and fantasy can cultivate international-mindedness about the issues that most connect us. As such, selected texts inspire (directly or indirectly) ecological consciousness, but also address morality in interspecies interactions and ethics in humanity's relationship with science and technology.

When readers begin to understand the changes society must collectively make to succeed, it is easy to become overwhelmed. As many environmental scientists cautioned, we have waited too long; the efforts humanity must make now seem herculean compared to the incremental steps international leaders were urged to take decades ago. Therefore, motivating learners to be positive contributors within a world community that envisions a better future

relies on careful consideration of individual responsibilities. Through various 21st Century Learning Initiatives related to the Common Core, public schools in the United States are experiencing relevant shifts in ideology, particularly as an extension of STEM (Science, Technology, Engineering, and Mathematics) or STEAM (Science, Technology, Engineering, Art, and Mathematics) programs intended to prepare students for competition in the world economy. Here and abroad, the International Baccalaureate Organization
promotes such a philosophy in primary, middle, and secondary levels of education. Encouraging a pedagogy toward global citizenry, their mission statement begins with the goal of "develop[ing] inquiring, knowledgeable and caring young people who help to create a better and more peaceful world through intercultural understanding and respect."[1] Indeed, the challenges we face as a global community presently demand not only "intercultural understanding and respect" among members but a united effort toward a "better and more peaceful world" that can accept responsibility for its past, organize action in its present, and envision progress for its future. A concrete approach to nurturing these global-minded goals could be developed in any school with ecocritical texts that specifically target universal concerns.

In *Global Citizenship and the University: Advancing Social Life and Relations in an Interdependent World*, Robert A. Rhoads—professor and founding director of the Globalization and Higher Education Research Center at the University of California—and Katalin Szelényi, assistant professor of higher education, explore the present state of internationally-minded education at the university level. They assert the importance of educational reform to cultivate cultural awareness and graduate global citizens prepared to participate in an international community. In reviewing the work of scholar Edward Said in *Culture and Imperialism*, the professors remind readers that the underlying cause of the "us" versus "them" mentality that pervades current culture in all parts of the world is education. Part of their project is to explore Said's conclusion that "educational institutions in general and universities in particular, [exist] as key sites for perpetuating particular notions of identity, nationalism, and citizenship," and to understand "the ways in which they foster the othering of a 'them.'"[2] Noting the changing demands of a rapidly advancing global society (due to technology, improved cultural relations, increased international mobility) they urge institutions to reevaluate their educational philosophy. Echoing sentiments of the IBO mission statement, the professors suggest a revision of what it means to be a global citizen:

> . . . it is not simply the geographic scope of one's actions as a citizen that constitute global citizenship, but rather it is the nature of one's understandings and the commitment to broader concerns that constitute global citizenship;

> thus, we see global citizenship as being marked by an understanding of global ties and connections and a commitment to the collective good.[3]

This "collective good" could not find a more accessible and relevant channeling than environmental activism. While current weather impacts are largely skewed based on geographic location and economic position, all reliable data point to long-term resource crises all will eventually face if we do not change course soon. Perhaps more importantly, our moral compass should not be appeased by following the status quo simply because consequences, for the moment, are happening elsewhere, to others. Moving forward, understanding the circumstances of climate change as a terrifyingly monumental but shared human experience is key to engineering effective change.

The authors' advocacy for global citizenry mainly addresses higher level institutions, but beginning the effort earlier could be more transformative. While their research does not specifically focus on the environment, it is an effective starting point as obvious common ground.

Though science tells us the climate-reality of our future is grim, imagination might help us (and our students) face it. In her introduction to *The Left Hand of Darkness*, Ursula LeGuin reflects on the uniqueness of science fiction writers' ability to do so: "Science fiction is often described, and even defined, as extrapolative. The science fiction writer is supposed to take a trend or phenomenon of the here-and-now, purify and intensify it for dramatic effect, and extend it into the future. 'If this goes on, this is what will happen.' A prediction is made."[4] Given this basic context, works of science fiction might serve an educational purpose by extending reading conversations on such predictions—whether they be social or technological or environmental. If education's goal is to prepare students as real-world problem solvers who understand the complexities of living, working, producing, and consuming in a global society, then educators' first task is to engage students in what those existing and potential problems are. The efforts of authors who offer those dilemmas—real, speculative, imagined—through stories could initiate conversations about why they matter or how they differ in various settings. Fiction is most up to the task of such engagement because, as LeGuin notes of reading a novel, "when we're done with it, we may find—if it's a good novel—that we're a bit different from what we were before we read it, that we have been changed a little, as if by having met a new fact, crossed a street we never crossed before. But it's very hard to *say* just what we learned, how we were changed."[5] Through judicious literary choices, educators can stimulate this change beyond the classroom. Science and speculative fiction offer a safe zone between our reality today and the consequences for our future; students need to see the time between present and projected realities as their opportunity for action and thus hope.

For middle or high school readers, fantasy may be an appropriate place to start because it provides a more comforting distance, especially in the form of otherworldly human or animal archetypes, while still prompting important reflection. Consider, for example, J.R.R. Tolkien's many brilliant battle scenes in *The Hobbit* and *The Lord of the Rings* trilogy, which Peter Jackson has directed in vivid screen adaptations. Perhaps one of the most moving is in *The Two Towers* when the Ents, ancient trees who have refused to participate in Middle Earth's war, seek revenge on the forces of Saruman, a symbolic embodiment of Industrial Revolution. Evil takes form as Saruman manufactures an army of creatures to defeat all others. When the battle has nearly ended in defeat, forces for good are outnumbered and no help appears on the horizon, it seems certain "the world of men will fall."[6] While the conflict is shaped by an archetypical good against evil paradigm of storytelling, readers might interpret the ensuing chaos as the more complicated clash between industrial progress and the preservation of nature—manufactured versus natural, Taker fighting Leaver. Even younger audiences seem to understand that who we root for in this epic battle speaks to who we are and for what we stand. But while most of us are not cheering for Saruman's army, are we cognizant of what that army represents in our daily decisions and lives, the ways we may be unconsciously supporting its fearsome strength? Attempting to answer questions like this connects literature to life with transformative effect.

It is difficult not to be moved in support of nature, for example, when Treebeard and endearing hobbits Merry and Pippin discover what remains of what was once a huge, old forest, their tree friends burned to ash. A potential interdisciplinary resource for delving into this scene with advanced science students is the recently published *The Hidden Life of Trees: What They Feel, How They Communicate—Discoveries from a Secret World.* by Peter Wohlleben. The insights Wohlleben shares would also enhance students' analysis of writing craft and purpose as they consider Tolkien's detailed presentations of the ecology of Middle Earth. In the novel, devastating deforestation symbolizes a turning point in the relations between humanity (represented at least in part by the power-driven wizard Saruman and his symbolic war beasts) and nature. As in life, the moral failure of those in power threatens all those who are not. Representing another side of humanity, however, is the courage of hobbits, dwarves, elves, and men. We might ask students what inherent knowledge drives empathy for their cause over the other; if our hegemonic cultural mind-set values power and strength, why are we standing with the underdogs? As Tolkien so epically conveys, the fight for good is inextricably connected with preserving the natural world. When the Ents march on Saruman to destroy what he has manufactured through destruction of the natural (both literally and symbolically), they are truly a force for nature that ultimately helps defeat the army of darkness. Appropriately, the battle fought to save Middle Earth is won through reconciliatory collaboration of all those who wish to live upon it. Our relief at this outcome invites readers to reflect on our own actions: in our real-world confrontations, which side are we supporting?

As Tolkien so wisely cautioned, the consequences of human behaviors—driven by a Taker philosophy of materialism and industrial progress—are far too dangerous, demanding what Outdoor Philosophy Professor Kate Rawles refers to as "A Copernican Revolution in Ethics." In her essay, Rawles pinpoints climate change as "a symptom of the dominant worldview of industrialized societies, a worldview that is profoundly flawed and that advocates and legitimizes ways of life and social structures that are, in a very literal sense, unsustainable."[7] Part of this flawed worldview, which is rapidly spreading to developing countries, "is its presentation of ethics as a purely interhuman affair" that "puts our own species alone at the center of the ethical universe and allocates value to all other species solely in relation to our own needs and interests, the ethical equivalent of the belief that the sun revolves around the Earth."[8] Rawles laments the failure of education to promote awareness about the issues of climate change or to engage young people in discussions regarding the ethics of anthropocentric perspectives that drive us. She emphasizes objectives educators might adopt as we consider a more authentic pedagogy: "The challenge is to radically reassess what is currently meant by 'success,' 'development,' and 'progress.'"[9] In its place she suggests educating toward creating more meaning in how we spend our time and define our interactions. As literature from all genres and of all cultures can provide contexts through which to focus such considerations, we can learn from stories like Tolkien's that who we cheer for matters.

There is a more deeply embedded ecopsychological rationale for taking sides in this metaphorical war, including real implications on our psyche if we fight for the wrong one. In defining progress toward the goal Rawles outlines, consider Ursula LeGuin's short story *The Ones Who Walk Away from Omelas,* in which readers are offered intriguing and plentiful details of a place so joyous the narrator admits (or more appropriately, feigns) difficulty in conveying it: "But I wish I could describe it better. I wish I could convince you. Omelas sounds in my words like a city in a fairy tale, long ago and far away, once upon a time."[10] To persuade disbelieving readers of this utopia, the narrator continues to describe the city's scapegoat—a young child isolated in a closet space, denied human contact or sunlight, and kept alive on "a half bowl of cornmeal and grease a day."[11] It is this child's misery, the narrator explains, that allows for the joy experienced by every other inhabitant of Omelas: "If the child were brought up into the sunlight out of that vile place, if it were cleaned and fed and comforted, that would be a good thing, indeed; but if it were done, in that day and hour all the prosperity and beauty and delight of Omelas would wither and be destroyed. Those are the terms."[12] Though not necessarily specific to ecological concerns, the archetypal scapegoat in this "city of joy" offers readers multiple metaphors for the unfortunates of their own social settings. Awareness comes from inside the individual, but educators can present situations or contexts that motivate discussion about real scapegoats that underlie our perceived safety, happiness, and wealth.

Invariably students are affected by the description of the child and certainly no reader would admit to wanting to remain in Omelas and "[face] this terrible paradox."[13] In fact, it is usually this point in the story when students demand to know why the people who come to see the child do not leave their city, which prompts LeGuin to provide for those reactions: "at times one of the adolescent girls or boys who go see the child does not go home to weep or rage, does not, in fact, go home at all. Sometimes also a man or a woman much older falls silent for a day or two, then leaves home . . . They keep walking, and walk straight out of the city of Omelas, through the beautiful gates."[14]

At these moments—when some students are wondering where the characters go when they leave the city—educators could enhance LeGuin's message by delving into the importance of consumer awareness. Depending on cultural background and geographical location, applicable references might range from the extraction processes of minerals used in cell phones, to the manufacturing of clothes and shoes, or the harvesting of agricultural and seafood products. Since so many "conveniences" have as scapegoats the children of various parts of the world, we can purposefully engage similar aged students in discussions of such realities—not as a means of creating guilt or sadness but as a path to responsible awareness and potential activism. Reactions to LeGuin's short story are powerful especially when paired with real world anecdotes from what otherwise may be anonymous places found on the labels of their clothing, shoes, food, or electronics. As a resource for defining what it means to be a Taker or Leaver culture in the global community, we owe young people the opportunity to decide for themselves whether to stay or leave.

Perhaps a subtler way of introducing awareness that inspires meaningful interactions is found in the popularly available platform of J. K. Rowling's *Harry Potter* series, appropriate for middle or early high school students. With varying degrees of seriousness, from the more light-hearted nature of magical adventure in the first pieces to the much more sinister exploration of the forces of good and evil in the later titles, Rowling's fantasy setting offers ample opportunities to discuss real world concerns. In an essay titled "Magic, Science, and the Ethics of Technology," contributing faculty authors Benjamin L Bruxvoort Lipscomb and W. Christopher Stewart offer this connection between Harry Potter's world of magic and real-world application for readers: "If, as we think, the use and abuse of magic in Rowling's books is a close analogue for the use and abuse of applied science—technology—in our world, then the ethical judgments her heroes make concerning magic may have something to teach us about an appropriate ethics of applied science."[15] Certainly, they affirm, "in Harry Potter's world, magic is as effective and pervasive a means of negotiating and manipulating one's surroundings as applied science has become in ours,"[16] so it is not a stretch to consider how the "indiscriminate" and "improper" use of magic presents similar problems for the characters of Harry's world as the abuse of technology does or can in ours.[17] As Lipscomb and Stewart helpfully point out,

> It is as a result of magic that the Giants are an endangered species (OP, p. 427), and what Giants hate most about witches and wizards is their magic (OP, p. 431). To cope with such problems, various regulatory agencies exist, such as the "Improper Use of Magic Office," the "Broom Regulatory Control," the "Apparition Test Center," and the "Department of Regulation and Control of Magical Creatures."[18]

Thus, what Rowling achieves in the conflicts of Hogwarts and its magical world is a familiar caution: despite its many conveniences, magic—like science—brings consequences, but it is not necessarily the witches and wizards (i.e., the scientists) who suffer them. It is the others (Giants and Muggles) in Harry's world who pay the price, while in ours it is other species, less dominant human cultures, the planet itself. Illuminating this discrepancy for students might include a discussion of how improved medical technologies have contributed to longer lifespans in developed countries even while malnutrition, starvation, and the lack of clean drinking water claim the lives of children all over the world. We need innovation, but we also must prepare young people for the basics of managing even more limited resources in their future.

What Libscomb and Stewart conclude in their review of Rowling's work as an educational resource is an interesting commentary we might apply to our own ethics in science and technology:

> Rowling's ethic suggests the importance of limiting oneself, not just to 'good application,' but to the use of certain kinds of powers over against others, based primarily on the attitude toward the world that these powers encourage. From such a willingness on the part of individuals ought naturally to flow a variety of self-imposed restrictions adopted by entire communities. Such self-restrictions will be part of the fabric of carefully structured educational programs, ones that include as a central component 'technology education.'[19]

Significantly, they do not mean the type of technology education students experience today, but rather one that instructs in "how our 'special powers' use us,"[20] one that ensures we understand the ways myriad advancements "encourage us to think about and treat our surroundings, and in how to make choices among possible uses of them."[21] As in Hogwarts School of Magic, we should not only be teaching our students how to use this gift but instructing them in the more important questions of to what extent and to what purpose.

More traditionally studied as a classic of the science fiction genre, Mary Shelley's *Frankenstein* turns students' attention to how the technologies we employ affect us and those around us. After Frankenstein has agreed to "create a female" for the "fiend"[22] that was the outcome of the scientist's ambitious intellectual endeavor, he finds himself alone in the wilderness, exclaiming "'Oh! stars, and clouds, and winds, ye are all about to mock me: if ye really pity me, crush sensation and memory; let me be as nought; but if not, depart, depart, and leave me in darkness.'"[23] Frankenstein's fear of the fiend's threat to harm his loved ones leads to his consent, but this solitary moment in nature will resurface later in the novel when Frankenstein realizes his debt to society. As he himself narrates to his listener—a man equally determined in his maritime expedition—"These were wild and miserable thoughts; but I cannot describe to you how the eternal twinkling of the stars weighed upon me, and how I listened to every blast of wind as if it were a dull ugly siroc on its way to consume me."[24] Significantly, the full weight of this moment strikes Frankenstein during his self-imposed isolation (he has even sent Clerval, his best friend and moral foil, away at this point) to create the fiend's female companion. In the reflection that follows, we see a scientist who measures what he can create against the individual and social consequences of doing so:

> Had I right, for my own benefit, to inflict this curse upon everlasting generations? I had before been moved by the sophisms of the being I had created; I had been struck senseless by his fiendish threats: but now, for the first time, the wickedness of my promise burst upon me; I shuddered to think that future ages might curse me as their pest, whose selfishness had not hesitated to buy its own peace at the price, perhaps, of the existence of the whole human race.[25]

Though Shelley places the burden of decision firmly on the scientist, readers are prompted here to consider the consequences if Frankenstein had not "trembling with passion, tor[n] to pieces the thing on which [he] was engaged."[26] Indeed, if readers sympathize at all with what Frankenstein stands to lose in this action, or pity the "wretch [who] saw [him] destroy the creature on whose future existence he depended on for happiness,"[27] we may imagine an alternative ending more true to dynamics of human nature—that out of selfishness or even empathy, Frankenstein completes his task. The actual consequence of this creation, though speculated by Frankenstein, is of course unknown. However, Shelley directs her readers' attention to Frankenstein's realization of potential consequence: "Had I right, for my own benefit, to inflict this curse upon everlasting generations?"[28] Regarding scientific endeavors as well as environmental abuses, this question holds profound relevance today.

Fictional possibilities often parallel present realities. Like Shelley's frame narrator of Frankenstein's tale, readers are cautioned to reflect upon perceived achievements by contemplating seriously the ambitions that drive them. This text, like so many others of its genre, can contribute in meaningful ways to developing scientific or technological courses that evaluate humanity's advancing capabilities within an ethical framework. The archetypes of Frankenstein as determined scientist and his creation as pitiable monster can serve as metaphor to any number of contemporary scientific ambitions—from machines that perform human tasks to genetic engineering of plants and animals (including humans).

Interestingly, Shelley's creature also challenges readers to confront humans' treatment of the Other—specifically himself but also cultural Others. In his determined plea to Frankenstein that he create a female to alleviate his loneliness, the creature painfully reminds his creator of the despair he and his "kind" have inflicted upon him: "If any being felt emotions of benevolence towards me, I should return them an hundred and an hundred fold: for that one creature's sake, I would make peace with the whole kind!"[29] As observed throughout the tale, humanity's constant inhumanity toward Frankenstein's "monster" is the cause of his present rage. Readers' shame at this recognition resurfaces when the creature learns about human history, particularly of conquest and slavery, through readings at the De Lacey home. Professor John Bugg's reading of the text in "'Master of the Language': Education and Exile in Mary Shelley's *Frankenstein,*" suggests "the Creature has difficulty understanding how human relationships could be based upon systematized racial and cultural violence: 'These wonderful narrations inspired me with strange feelings. Was man, indeed, at once so powerful, so virtuous, and magnificent, yet so vicious and base?' (p. 94)."[30] Gaining awareness of our own role in this inhumanity involves examining how we exercise personal freedom at the expense of others—from intentional negative interactions with peers to unintentional ignorance in our consumer habits.

A more direct critique of humanity's planetary stewardship as it regards other animals is David Brin's novel *Earth,* which imagines the grim future ensured if we continue destructive overuse of resources and ignore scientific climate warnings. One of the main characters, Jen, is a scientist trying to save Earth's environment and the many endangered species in it. An archetypal elder, mentor, and guide, readers learn that Jen "approve[s] of the way everyone marched and spoke out and wrote letters these days, passing laws and designing technologies to 'save the Earth' from twentieth-century errors."[31] As an activist Jen is involved in an ark project where scientists have created "enclosed, artificial habitats" with "meticulously regulated environments"[32] designed to preserve the many species near extinction due to humanity's abuse of the environment. Teachers could analyze this passage in light of current extinction and endangerment records or pair it with any number of subject-specific related nonfiction.

Take, for example, Alan Weisman's insight regarding Paul Martin's *Twilight of the Mammoths: Ice Age Extinctions and the Rewilding of America:*

> . . . [the book] conclude[s] with a plea that his accounting of the Pleistocene mega-massacre be a cautionary lesson that stops us from perpetuating another that would be far more devastating. The matter is more complicated than a killer instinct that never relents until another species is gone. It involves acquisitive instincts that also can't tell when to stop, until something we never intended to harm is fatally deprived of something it needs. We don't actually have to shoot song birds to remove them from the sky. Take away enough of their homes or sustenance, and they fall dead on their own.[33]

A comparison of the passages proves relevant by fostering interdisciplinary study of potential global extinctions, particularly those that students may have concrete opportunities to combat. Using Weisman's assessment about song birds in a mathematics course, for example, students could evaluate current statistics regarding endangered bird species. A history or science class might take up the challenge of contextualizing numbers with an analysis of short- and long-term causes of species' decline. An English class could report on the impact of Rachel Carson's *Silent Spring,* specifically her stylistic choice to introduce science through fiction, thereby compelling a potentially larger readership. Imagine if these students shared their findings with peers in a woodworking class or the community 4H or Scouts clubs, encouraging them to build birdhouses for local use or conduct community awareness presentations.

A few years ago, one of my English Honors seniors decided to focus her culminating graduation project on the endangerment of Eastern Bluebirds. She researched the issue, educated her class in a formal presentation of findings, worked with a local craftsman to build a size-appropriate birdhouse, and gave the final product to her grandfather—an avid bird watcher. This is an individual step inspired by pursuits to fulfill an open-ended project requirement, but thirty of her peers now know more about the dangers these birds face than they did before. I've since put a bluebird house in my own backyard as have several of her peers. After her presentation, many students noted they had no idea the birds were so threatened or that they could act so easily to improve their plight. Some began exploring the status of other local bird species, expressing genuine concern for how they could make a change. Once they knew how they could act on the issue, they were eager to help.

Beyond the ethical and practical dilemmas surrounding the preservation of endangered habitats and species, Brin deepens our introspection on human nature. In an enlightening moment of the text, he develops a cognitive rationale for humanity's stewardship of the earth by likening our evolutionary history to other animals. Nelson, a stereotypical "loser" who works in an ark project collecting for inspection samples of baboon feces, finds himself in a dangerous predicament when female members of the baboon ark he is in have joined forces against a lower female carrying her new infant. As Nelson realizes through his observations of their behavior and the singled-out female who seeks Nelson's support, the baboon females do not intend to harm the other female but to kill her infant. In a struggle to understand what he perceives as the baboons' cruel motivations, which were "so awful" they reminded him "of *human* nastiness,"[34] Nelson is torn between self-preservation—against the male baboons who threaten his own safety—and coming to the aid of the targeted female baboon and her infant:

> *We were all like this, once,* he realized, feeling the weight of his makeshift cudgel.
>
> *White, black, yellow . . . men, women . . . our ancestors all shared this, long ago . . .*
>
> *Back when Africa was new . . .*
>
> Human beings had changed the world, for well and ill. Would their efforts now save what was left? Nelson couldn't begin to guess.
>
> All he knew for sure was that for the first time he *cared.*[35]

No mere conflict of human compassion but a deliberately effective commentary on evolutionary behavior, this scene offers the opportunity to compare engaging speculative fiction with hard science on humanity's shared history with other mammals. While illuminating further the biological impetuses behind the freedoms we enact at the expense of others, Nelson's perspective also connects us to international counterparts with similar but often competing motivations.

Another compelling character in *Earth* representing these connections is Alex Lustig, whose ambition recalls that of Shelley's Victor Frankenstein; curiosity leads this scientist to uncover secrets of space and gravity to create a black hole, which he thought might be used toward energy sourcing. Like Frankenstein, Lustig's success quickly spirals out of control, providing the main premise for Brin's novel—an energy that is literally eating away Earth's core as Lustig's team searches desperately for a way to stop or control it. Just as Frankenstein came to realize, Brin's novel shows us—through the episodes of multiple other characters who experience environmental dilemmas related to Lustig's hole—how interconnected we are, how our choices in one part of the world definitively impact others. This might remind readers of a popular allusion to the butterfly's wing concept (as in the film *The Butterfly Effect,* in which seemingly insignificant actions far removed from the butterfly mushroom into cataclysms).

As environmentalist Bill McKibben explains in *The End of Nature*, our perspectives on ecological problems today must therefore involve both local and global awareness for effective planning:

> Our local problem here in the Adirondacks—acid rain—has its cause in Ohio and Kentucky. And now, as the climate warms, our local problem—the death of trees—starts to have its causes everywhere. Everywhere. A factory in Japan is as deadly as a burning rain forest in Brazil, a Communist coal mine in Romania, a capitalist utility in West Virginia. Or as the blue 1981 Honda parked in the driveway twenty feet from where I sit, or as the wood stove warming my back.[36]

Readers of Brin's *Earth* might therefore analyze the ecological disasters caused by Lustig's creation, as well as the general abuse of the planet Brin consistently references, to gain appreciation for how different freedoms link to current environmental disasters such as earthquakes, tsunamis, hurricanes, floods, droughts.

Other contemporary speculative fiction examining these issues include Margaret Atwood's *Oryx and Crake*, *The Year of the Flood*, and most recently, *MaddAddam*. This trilogy presents readers with a complex future whose settings reflect uncomfortable outcomes of current human behaviors. For example, readers' initial connection to the character of Snowman as the only human is soon conflicted by descriptions of Snowman's life. Isolated by his differences from the Crakers (his former friend Crake's perfected race of humans), we come to understand that while this world is not our own, it is most definitely an eventual product of it. In Atwood's setting, readers learn what the demands of contemporary society might achieve: a pill for every physical, emotional, or mental complaint; a complete renewal of tired, aging skin or organs (now grown in the bodies of genetically modified pigs called "pigoons"); a food supply that is not reliant on actual animals or land but scientific labs; specialty entertainment and pets to meet our every whim; security from all who wish to steal our secrets or threaten our lifestyle. It is a world where many real scientific hopes and possibilities manifest themselves in disturbing ways. Responses toward these popularized advances is mixed depending on the characters we meet. As in capitalist society, those who stand to benefit the most support the changes through financial investment (as with Crake's uncle's stock share in the Happicuppa bean) or by joining company monopolies such as HethWyzer or OrganInc (like Snowman's father). Others, like Snowman's or Crake's mothers, choose escape or isolation or protest—all of which result in punitive consequences or mysterious deaths.

Atwood critiques big business' ability to trample fair trade practices through the lens of the Happicuppa bean which has been modified to produce coffee in large quantities: "coffee could be grown on huge plantations and harvested with machines. This threw the small growers out of business and reduced both them and their labourers to starvation-level poverty."[37] A commentary on human violence follows the broadcasts about society's resistance to the Happicuppa bean: "There were protests and demonstrations, with tear gas and shooting and bludgeoning; then more protests, more demonstrations, more tear gas, more shooting, more bludgeoning."[38] As students will quickly discern, Atwood explores many legitimate issues in this text that are patterned on a mentality of "us" versus "them"—whether "them" refers to another species or to other humans less fortunate than "us." A character like Snowman/Jimmy is one with whom we can relate and therefore empathize, so it is a natural extension for readers to feel as Jimmy does about the injustices Atwood depicts. This commiseration will be important toward the conclusion of *MaddAddam* when Atwood presents readers with a collective effort by humans, Crakers, and pigoons to defeat the evil Painballers.

In the sister novel to *Oryx and Crake—The Year of the Flood*—Atwood examines other international concerns. Adam One's prayer with God's Gardeners (a group of surviving humans who seek a return to nature and spirituality) reveals an ecological critique readers cannot possibly deny given the present status of Earth's water ecosystems: "Let love and aid be brought to the Sea Creatures in their present peril and great suffering; which has come to them through the warming of the Sea, and through the dragging of nets and hooks along the bottom of it, and through the slaughtering of all within it."[39] In addition to the effects of climate change and overconsumption, students could familiarize themselves with a purely wasteful cause of sea life destruction: plastic trash. Alan Weisman's *The World Without Us*, published in 2007, includes a chapter called "Polymers Are Forever" in which he explores the extent of plastics found in the world's oceans. It provides a sad but powerful accompaniment to the Gardener's prayer. A descriptive passage that may intrigue students is one in which Weisman summarizes experiences of marine biologist Richard Thompson on a quest to assess the amounts and effects of plastic discarded into the sea:

> He knew the terrible tales of sea otters choking on polyethelyne rings from beer six-packs; of swans and gulls strangled by nylon nets and fishing lines; of a green sea turtle in Hawaii dead with a pocket comb, a foot of nylon rope, and a toy truck wheel lodged in its gut. His personal worst was a study on fulmar carcasses washed ashore on North Sea coastlines. Ninety-five percent had plastic in their stomachs—an average of 44 pieces per bird. A proportional amount in a human being would weigh nearly five pounds.[40]

Weisman's narrative makes statistics vivid and places blame squarely on the shoulders of human beings. His insistence could motivate some to learn more about trash pollution. Elective classes such as art might investigate the wonderful work being done by various activist artists bringing attention to impacts of equally spiraling consumer cultures through creative portrayals of the "plastic ocean" or "trash collector carts." Students interested in engineering might explore the instruments being developed to extract larger amounts of plastic from rivers, lakes, and oceans than can be collected through simple human effort. Education that includes awareness about how humans use, reuse, recycle, and discard the products they demand puts our best potential problem-solvers at the helm.

Another interaction in Atwood's setting recalls Frankenstein's grief-stricken motivation to create an improved, immortal version of humans. Two gardeners—Amanda and Glenn—converse about their friend Pilar's illness, revealing the origins of Snowman's companions, the Crakers. They have been designed, we learn, by the character of Glenn, who Jimmy refers to as Crake, through genetic engineering of human cells:

> Going back to the Tree of life, Glenn walked along beside us. 'There's something really wrong with her,' said Amanda. 'Right?'
>
> 'Illness is a design fault,' said the boy. 'It could be corrected.'
>
> 'So if you were making the world, you'd make it better?' I said. Better than God, was what I meant. . . .
>
> 'Yes,' he said. 'As a matter of fact, I would.'[41]

To create this new and "better" version of humans, Glenn/Crake designs a pill that annihilates almost all humanity including himself. As readers learn in *Oryx and Crake*, Crake has left Snowman/Jimmy as the guardian of this special species—the Crakers—who he intends will repopulate the earth and take better care of it than their predecessors. Snowman struggles with their improved but often unrecognizable attributes—such as their perfect physical forms, but especially their innocence of all things vilely human (greed, violence, rape, torture). As Atwood comments in her 2011 commentary *In Other Worlds: SF and the Human Imagination*:

> They are designer people. But anyone who engages in such design—as we are now doing—has to ask, How far can humans go in the alteration department before those altered cease to be human? Which of our features are at the core of our being? What a piece of work is man, and now that we ourselves can be the workmen, what pieces of this work shall we drop off?[42]

She further notes that the "Chickie Nobs solution" of her story—"chicken objects modified so they grow multiple legs, wings, and breasts" thereby "solving a problem for animal rights workers" because they "have no heads"—"has made giant strides: lab grown meat is now a reality" (since publication of *Oryx and Crake* in 2003).[43] This merging of science fiction and today's innovations is chilling but makes our engagement with such a text amazingly relevant.

As an allusive reminder of the dystopic societies created by Orwell (*1984*) and Huxley (*Brave New World*), Atwood notes that her God's Gardeners "avoid high-tech communications devices such as cell phones and computers on the grounds that they can be used to spy on you—which is entirely true."[44] This aspect of the story returns us to the topic of improving technology education in our schools through supplemental nonfiction texts such as David Brin's *The Transparent Society: Will Technology Force Us to Choose Between Privacy and Freedom?* In it, Brin documents the increasingly transparent nature of our lives due to technological advances (such as miniature video cameras and tracking devices in mobile phones). Referencing actual as well as hypothetical scenarios from different parts of the world, his work delivers much material for debate. From infringement of privacy rights to individual versus governmental use to promises of impeccable security, readers are asked to weigh the future impacts of technology on both private and social life. Posing a probing question for the emerging environmentalist or technologist, Brin asks "what will happen when personal computers become so cheap that citizens of the poorest Third World nations will have readier access to data than food or clean water?"[45]

For struggling high school readers or middle school students, we might focus this question on a novel such as Cameron Stracher's *The Water Wars*. In this future setting technology such as Brin describes is well-advanced but humanity's most basic of resources—water—is in very short supply. Fresh untreated water is almost impossible to access if you are not incredibly wealthy (like the government officials) or resourceful (like the pirates) leaving different U.S. "republics" and other countries of the world to compete over remaining supplies. The depletion of resources has led to a child labor dilemma similar to that of diamond or mineral mining. The narrator and main character—a young girl named Vera—describes this when she and her brother Will have been sold to a water drilling operation: "Most striking, however, were the children: thousands of them scrambling over the piles of dirt or shimmying down into crevices between rocks. Deep in the canyon bottom, they scurried from drill hole to drill hole, emerging into the gloom like colonies of insects."[46] Like Vera, readers are disturbed by the depiction of these desperate children. As is the case with Snowman's description in Atwood's story, Stracher's young adult readers are more alert to Vera and her brother's circumstance because her characterizations have carefully established an empathic connection. They could be them.

As part of their quest to save their friend Kai—who is labeled as a water diviner for his ability to identify remaining subterranean aquifers—Vera and

Will learn about companies responsible for the plight of these children as well as for the constant wars waged over water:

> Across the world there were other companies like Bluewater, poisoning the sea so people could turn on the tap without worrying about the consequences. To gain access to water, the lower republics were fighting a war against Canada and the Arctic Archipelago. Across the globe there were other wars between Japan and China, between Australia and New Zealand, between Argentina and the Kingdom of Brazil.[47]

Their experience at Bluewater helps to shape this environmental commentary into a global context of "us" versus "them" reactions that threaten to destroy the planet. Vera's reflections outside a monument to Bluewater—a company that achieves its wealth and power through desalination of ocean water—revisit the butterfly effect on different terms: ". . . the shiny surfaces hid corrosion and corruption. The entire edifice was a monument to ignorance. The truth was that butterflies could not disrupt an entire ecosystem simply by beating their wings. It took willful neglect and deliberate blindness, the refusal to see the obvious even as the land grew toxic before our eyes."[48] In addition to delving into details of our "willful neglect," Stracher's plot mirrors our own water wars. One contemporary example—the 2017 protests of the Dakota Access Pipeline in Standing Rock—relates equally to our treatment of the historical "Other" as it does the planet. As with Silko's *Ceremony*, the specifics of how this Taker-Leaver conflict unfolds will be extremely consequential both historically and environmentally. Still, discussions must not lead students down the path of guilt or hopelessness but rather toward a proactive, global mind-set. As young Vera claims, "I still held out hope that we could change our ways."[49] This hope echoes the fortitude demonstrated by citizens of the Standing Rock Sioux Nation and thousands of fellow protestors. Their determination in the face of all odds is at the crux of what must inspire a new generation of learners.

In defense of a genre so swiftly cut from reading lists when standards for mandated curricula interfere, education stakeholders might consider what Cultural Studies Professor Thomas Moylan calls the "empowering critical practice"[50] that popular science fiction offers students. In the first chapter of *Scraps of the Untainted Sky: Science Fiction, Utopia, Dystopia*, Moylan discusses how cognitive processing for science fiction readers is particularly challenging when given culturally conflicting archetypes. Reviewing Joanna Russ' utopian story "When It Changed," whose "distant planet" setting is populated only by "highly trained and resourceful professional women,"[51] Moylan comments on the difficulty readers initially face in trying to identify characters as female when readers' assumptions align with culturally defined male attributes: "They enter into recurring feedback loops that run from their hesitation over the manifest content of the text, into their own accumulated sense of that reality, and on to the emerging social paradigm that is absent of the page but active in the imagination."[52] Thus, Russ' story, as with texts like LeGuin's *The Left Hand of Darkness*, Gilman's *Herland*, Butler's *Wild Seed,* or Atwood's *The HandMaid's Tale,* offers readers a way not only to negotiate feminist issues and culturally determined gender stereotypes (even if only in their rereading because assumptions conflict with what is on the page) but an opportunity to explore how those issues and stereotypes manifest themselves in our cognitive interaction with texts. In other words, we can talk to our students not only about why they read but how. As Moylan comments of this reading process, it is a "complex engagement that enters into a dialectical negotiation of the historical tension between what was, what is, and what is coming to be."[53] Further, and particularly important as contributions toward goals of global citizenry, it offers us "the capability of elucidating the contradictions at work in a given social context (a given conjuncture) out of which the next moment of history (that which gets reified in the word *future*) is emerging."[54] Literature, as we may often tell our students, is thus a way of teaching us how we live as much as it teaches us how to live.

As an educational tool, literature can "mobil[ize] the cultural imagination. It can be part of the process of making the world critically 'legible' in a way that not only delivers pleasure and knowledge but also the joys of joining in the collective, historical work of bringing a more just and free society into being."[55] The rereading students must engage in to appreciate texts that manipulate stereotypical cultural archetypes is a beneficial exercise toward this larger goal of contextualizing any problem—environmental, political, social, or economic—in a framework of local as well as global responsibility. Given our position as shareholders in a rapidly declining ecosystem, educational approaches that engender this type of critical thinking should influence all content areas. Science fiction and fantasy provide a medium for exploring real world challenges while contemplating potential interventions and solutions. In the characters and conflicts of fictive dystopias, readers can better recognize the mistakes we are making and the lessons we must not ignore.

NOTES

1. Mission Statement, International Baccalaureate Organization, www.ibo.org/missionstatement.
2. Edward Said, *Culture and Imperialism* (New York: Harcourt Brace Jovanovich, 1993), 20, quoted in Robert A. Rhoads and Katalin Szelényi, *Global Citizenship and the University: Advancing Social Life and Relations in an Interdependent World* (Stanford: Stanford University Press, 2011), 6.
3. Robert A. Rhoads and Katalin Szelényi, *Global Citizenship and the University: Advancing Social Life and Relations in an Interdependent World* (Stanford: Stanford University Press, 2011), 27.
4. Ursula LeGuin, *The Left Hand of Darkness*, Introduction (New York: Walker and Company, 1969), vii.

5. Ibid., x.
6. *The Lord of the Rings: The Fellowship of the Ring*, directed by Peter Jackson (2001; USA: WarnerBrothers Pictures), DVD.
7. Kate Rawles, "A Copernican Revolution in Ethics," in *Moral Ground*, 88.
8. bid., 91.
9. Ibid., 94.
10. Ursula LeGuin, *The Ones Who Walk Away from Omelas*, http://www-rohan.sdsu.edu/faculty/dunnweb/rprnts.omelas.pdf, 1.
11. Ibid., 2.
12. Ibid.
13. Ibid.
14. Ibid., 3.
15. Benjamin L Bruxvoort Lipscomb and W. Christopher Stewart, "Magic, Science, and the Ethics of Technology" in *Harry Potter and Philosophy: If Aristotle Ran Hogwarts*, ed. David Baggett and Shawn E. Klein in *Popular Culture and Philosophy* 9, ed. William Irwin (Chicago and LaSalle: Open Court, 2004), 77.
16. Ibid., 79.
17. Ibid., 81.
18. Ibid.
19. Ibid., 91.
20. Ibid.
21. Ibid.
22. Mary Shelley, *Frankenstein with Related Readings* (New York: Glencoe McGraw-Hill, 2000), 106.
23. Ibid., 109.
24. Ibid.
25. Ibid., 122.
26. Ibid., 123.
27. Ibid.
28. Ibid., 122.
29. Ibid., 107.
30. John Bugg, ""Master of their Language": Education and Exile in Mary Shelley's Frankenstein," *Huntington Library Quarterly* 68, no. 4 (December 2005), 662, http://www.jstor.org/stable/10.1525/hlq.2005.68.4.655.
31. David Brin, *Earth* (New York: Bantom Books, 1990), 14.
32. Ibid., 12.
33. Alan Weisman, *The World Without Us* (New York: Picador, 2008), 6 –7.
34. Brin, *Earth*, 94.
35. Ibid., 85.
36. Bill McKibben, *The End of Nature* (New York: Random House, 2006), 32.
37. Margaret Atwood, *Oryx and Crake* (New York: Random House, 2003), 179.
38. Ibid.
39. Margaret Atwood, *The Year of the Flood* (New York: Random House. 2009), 273.
40. Alan Weisman, 116.
41. Margaret Atwood, *The Year of the Flood,* 204.
42. Margaret Atwood, *In Other Worlds: SF and the Human Imagination* (New York: Anchor Books, 2011), 91.
43. Ibid., 92.
44. Ibid., 93.
45. David Brin, *The Transparent Society: Will Technology Force Us to Choose Between Privacy and Freedom?* (Cambridge, MA: Perseus Books Group, 1998), 41.
46. Cameron Stracher, *The Water Wars (*Naperville: Sourcebooks Fire, 2011), 135.
47. Ibid., 186.
48. Ibid., 200.
49. Ibid.

50. Thomas Moylan, *Scraps of the Untainted Sky: Science Fiction, Utopia, Dystopia* (Boulder: Westview Press, 2000), 26.
51. Ibid., 9.
52. Ibid., 11.
53. Ibid., 25.
54. Ibid., 26.
55. Ibid., 28.

Chapter Four

Let's Share the Table

Building Ecoliterate Communities

In the words of Raymond Carver, food is "A Small, Good Thing." Celebrating its healing, communal nature, Carver's short story of this title poignantly documents the heartbreak of family tragedy, the long process that is grief, and the surprising comforts compassionate strangers can provide. He demonstrates not only the therapeutic influence of food but also its ability to redeem and unite us in the most challenging of moments. As we now face our most daunting challenge—what David Orr refers to as "dangerous years"[1] —this chapter explores food as a simple yet essential resource for restoring humanity to a shared cultural mind-set that honors our best selves.

As I share in the resource *Teaching Climate Change to Adolescents: Reading, Writing, and Making a Difference*, high school seniors played a part in teaching me this during the closing discussion of a nonfiction food ethics unit that included upper level science fiction texts. Complementing our study of relevant agricultural factors (including health, nutrition, environment, production, distribution) were climate change or cli-fi narratives such as Margaret Atwood's *Oryx and Crake*, Paolo Bacigalupi's *The Windup Girl,* Cormac McCarthy's *The Road*, and Jim Crace's *The Pesthouse*, among others. In literature groups, students wavered in their support or disapproval of characters, excitedly analyzing and reporting on their conflicts, while as a class we researched issues, argued viewpoints, and posed more questions. In our final group talk, Emily—a bright student who seemed to find her voice now out of a passion for healthful living—pointedly asked: "So, what do we *do* with all this information now that we know it?" Her tone was serious, the inquiry genuine, immediately drawing the attention of the thirty-two other teenagers in the room. As if on cue, every student in our circle turned to me. The anticipatory silence of their sudden concentration told me they were ready to transfer what we had studied in class into real changes in their lives. Now they were waiting for me to tell them how to do it. The moment was gratifying and terrifying at once: literature and life had come together in a most transformative potential but education had not prepared them to make the transformation.[2]

Indeed, it had not prepared me either. After a long pause, I admitted there were no easy answers, no quick fixes, but many, many steps we could take toward living and eating better. I told them about my own journey of learning to care about food by investigating the myriad ethical, health, environmental and social justice issues surrounding it. I shared some of the steps I have taken, laughing with them about my blunders. Then, I put it back to the group: "So, what do you think, what can we do?" Gradually, a few raised their hands to explain steps they had taken to change their eating and shopping habits (after consulting with their families about what was now "off-limits" at the grocery store). Others excitedly began swapping stories of how they had already been supporting local farmers, without realizing how important it was. Ben, who had humorous stories of helping his grandparents with their chickens and had already been selling eggs to me and a few other teachers, picked up some new customers. He also shared his concerns about chemical pesticides and that he planned to do his senior project on hydroponics. We were having serious fun. The practical, locally available steps we discussed demonstrated for these soon-to-be graduates how they might extend the empathy they felt for their characters' situations to bring about changes in their own. Fiction had exposed their unconscious roles as Takers, so they consciously decided to respond as Leavers.

As such a potentially powerful teaching lens, it is difficult to imagine why food is not already a significant part of curricula. While many schools across the United States have experimented with different aspects of environmental-based food education (such as engaging in local food movements, place-based education, and school garden projects), ecoliteracy has remained largely on the sidelines rather than as a central tenant for pedagogical reform. Food as a focus of education, particularly as part of reading, listening, analyzing, and telling stories, fosters the importance of place through awareness of local climate conditions and culinary traditions. As neutral common ground, food also provides a foundation for ongoing conversations about multiple interdisciplinary issues: water, land, and air quality; food production and distribution trends; health, illness, and wellness; scientific and technological applications; changing weather patterns; pesticides and chemical fertilizers; malnutrition, starvation, and poverty; the treatment of animals raised for food, among many others. The experience of sharing food creates opportunities for learning social mores or participating in food-related customs. Hands-on learning experiences with growing food help build a communal work ethic that invites each contributor to enjoy its bounty. Lessons in harvesting, preparing, eating, and composting food, can foster respect, gratitude, responsibility. Studies in food ethics help distinguish practices that perpetuate the competitive individualism of American culture from those that reflect compassion for each other, nonhuman animals, and Earth. Education initiatives that bring awareness to the impacts of our relationship with food could result in benefits that include improved nutrition, healthier environmental practices, conscientious consumer behavior, and more prosocial interactions.

Educators interested in exploring these benefits but perhaps reluctant to broach topics outside their curricular area might first consult Wendell Berry's *The Unsettling of America.* An incredibly helpful resource for understanding the importance of food on multiple levels, it begins with a telling reference to French philosopher Michel de Montaigne that recalls the Taker-Leaver paradigm: "Who so hath his mind on taking, hath it no more on what he hath taken."[3] In the first chapter, appropriately titled "The Ecological Crisis as a Crisis of Character," Berry illuminates humanity's role in remedying irresponsible behaviors toward the earth. As in Ishmael's lessons, readers learn to recognize that with knowledge comes free will: "Once our personal connection to what is wrong becomes clear, then we have to choose: we can go on as before, recognizing our dishonesty and living with it the best we can, or we can begin the effort to change the way we think and live."[4] Berry's acknowledgment of our free will is not focused on the broad shift from hunter-gatherer to agricultural societies but specifically on the consequences of American agriculture's transformation from family run farms to corporate-dominated factory operations. In *Made for Each Other: The Biology of the Human-Animal Bond*, Meg Daley Olmert explores this history in congruence with the biology of human-animal interactions and their resulting evolutionary benefits. In reference to agropastoral practices of the past, she notes that "the care of plants and animals once caused us to settle down, learn to live together, and think of ourselves as caretakers and citizens. For twelve thousand years, we sacrificed self-interests to the care of each other, our crops, and our animals."[5] In a society where "self-interests" have become paramount, it is important to take note of the cultural changes that allowed this shift in priorities from social to individual welfare. As both authors suggest, the reality of this transformation can be found in our unprecedented abandonment of family-based agriculture. Olmert summarizes it:

> In 1920 a third of all Americans—32 million of us—still farmed the land. By 1950 that number had slipped to 23 million. Forty years later it was down to 4.6 million—less than two percent—and a third of those farmers didn't even live on the land they were farming. By 1993 farmers were so rare that the United States Census Bureau stopped counting. The family farm was extinct.[6]

Our youth deserve to know this history as it has directly contributed to today's consumer-driven culture. As participants who now face the serious consequences of this culture, students are entitled to understand why farmers like Berry no longer recognize the practice today; they have a right to be informed about the agricultural products and practices that affect them. Understanding multiple interrelated consequences of this usurpation of family farming by industrial agribusiness is essential to addressing them.[7] One basic example is the influence this knowledge is having in the organic foods market. As more and more consumers learn about growing practices, the market has been experiencing an undeniable demand for organic options at competitive prices. Similarly, as more buyers learn about the importance of supporting local farming, some states are witnessing a surge in farmer's markets, community supported agriculture programs, and the provision of local goods in grocery stores. Smaller scale farming whether in our backyards, community gardens, and city rooftops has surged in popularity. Awareness can lead to a host of other positives from better understanding of nutritional labels to supporting fair trade practices to reconnecting with the sources of our sustenance.

By illustrating how the prevailing harmful ethos infiltrates all aspects of our lives, Berry's work provides a foundational philosophy for reforming it. Investigating how as a culture we have forgotten our kinship with the earth, forgotten what we have taken, Berry's message details how this continued willful forgetting has damaged our most precious cultural identities. Discussing education, government, marriage, personal responsibility, the free market, farming, heath care, the environment, and more, he considers these damages from the perspective of myth and metaphor, characterizing Takers' modern obsession with the machine:

> Once, the governing human metaphor was pastoral or agricultural, and it clarified, and so preserved in human care, the natural cycles of birth, growth, death, and decay. But modern humanity's governing metaphor is that of the machine. Having placed ourselves in charge of Creation, we began to mechanize both the Creation itself and our conception of it. We began to see the whole Creation merely as raw material, to be transformed by machines into a manufactured Paradise.[8]

As we saw in Quinn's novel, Berry scolds us for having squandered Leaver ideologies for Earth-abusing practices that serve our needs more cheaply and quickly. The cost of these practices, of course, manifests itself in more complex ways. It is evidenced, according to Berry, in a failing education system, hypocrisies in politics, the breakdown of American families, corruptions of free markets, eradication of farm land, an epidemic of preventable illnesses, disintegration of natural resources, and perhaps most blatantly, dismissal of individual responsibility.

If students have difficulty connecting the machine metaphor to these broad implications, educators might focus them on the more intimate effects they can recognize—from the absence of family dinners at the table to not knowing even our closest neighbors' names. As Berry references, this ethos of mechanization pervades every aspect of our lives—from birth to death. Just as life-sustaining food is a common ground for bringing awareness to complex topics, so is death. Especially for students who may not be able to see the more complex or abstract effects of the current ethos, literature can illustrate them in more concrete terms.

A personal essay in the *Moral Ground* collection offers, for example, an accessible lesson for explicating Berry's intimation that we disrupt even the "natural cycles of birth, growth, death, and decay." Carly Lettero's "Spray Glue Goes. Maggots Stay." focuses on how an anthropocentric worldview of "progress" follows us even in our deaths, separating us superficially from nature and its processes. The narrative achieves this by juxtaposing the ways humans treat a body after death with how nature does. Upon witnessing the hospital's procedural sanitizing of her grandfather's dead body, which includes "so much disposable crap" in the form of "hermetically sealed cotton swabs," "rubber gloves," "individually wrapped wipes and plastic padded sheets,"[9] Lettero recalls a very different memory of finding the body of a sea lion that had washed ashore on a coast of Oregon:

> Turkey vultures pecked into its stomach and chest, and black fluid flooded the holes. Closer still the holes were crawling with bugs. Maggots slithered through the black liquid and burrowed into the soft tissue. Flies swarmed above the body, landed, got swept down the beach by a gust of wind, and swarmed again. So much life spiraled out from this one dead animal. In stark contrast to my grandfather's death, nothing was wasted.[10]

This sensory description of nature at work sharply contrasts the mechanized approach to human burial she witnesses:

> [Two men] pulled a white bag of impermeable plastic over the casket and sealed the edges with spray glue. . . . They raised the whole package precariously toward the cold afternoon sky, slid it into a cement wall, and closed it in with a sheet of Styrofoam, a thin piece of marble, and four screws. Were bugs able to squeeze through the fortress, they would die of formaldehyde poisoning from the embalming fluid. This body entombed in a wall was no longer my grandfather. It had become toxic waste.[11]

The stark differences observed in these processes offer a clear and personal context for Berry's machine metaphor. Lettero highlights in a simple comparison the drastic measures we employ to position humans above all else in nature, to view ourselves as distinct from rather than as inclusive of other species or the earth. It is this ethic that blindly drives our life pursuits and, as this provocative narrative illustrates, which falsely positions us in artificial opposition to natural processes. This opens the possibility of exploring different cultural practices regarding death, including whether we should consider ecologically responsible alternatives to ours. On a reflective note that challenges readers to translate her disturbing observation of death into purposeful action in life, Lettero contrasts her grandfather's generation with her own, highlighting unique obstacles faced by each and prompting hers to act:

> But my grandfather, and the previous generations before him did lots of things that were not easy. They fought two world wars, survived the Great Depression, and sent a man to the moon. Can't my generation stop throwing stuff away, curb greenhouse gas emissions, and ban pollution? Thomas Jefferson once said that every generation needs a revolution. I believe this is ours.[12]

We should consider this "revolution" not only in our immediate behavior toward the earth but in improving our interactions with other species and each other.[13]

As Berry ultimately uncovers, the shift in the way human beings view Earth—from graciousness in its bounty to shameless utility and disrespect of its limits—reaches beyond physical consequences into the emotional and spiritual:

> By means of the machine metaphor we have eliminated any fear or awe or reverence or humility or delight or joy that might have restrained us in our use of the world. . . . Our "success" is a catastrophic demonstration of our failure.[14]

Devotion to the machine (and all that it literally and symbolically represents) has replaced our role as stewards of the earth with the goal of being masters of it. This sinister vantage point suggests that most of us have become Takers without an appreciation of what it means to be a Leaver. Instead of perpetuating existing discourses regarding our natural resources—maximize usage to meet future production demands—Berry argues that "the only possible guarantee of the future is responsible behavior in the present. When supposed future needs are used to justify misbehavior in the present, as is the tendency with us, then we are both perverting the present and diminishing the future."[15] Shocking disregard for natural life cycles of plants, animals, and farm land illustrate the utility of this "future" rationale in agriculture today; similarly, demands for alternative energy sources fuel current debates regarding the safety of hydrofracking to retrieve natural gas reserves. Despite well-documented scientific evidence cautioning against the chimera of progress in both areas, appeals to the future seem, ironically, to convincingly justify present abuses. Through intentional curricular changes, educators can help students investigate additional examples that will directly impact their future. If literature can help young people recognize flaws in this sense of superior control humans have imagined for themselves, it can guide them toward a more empathic social narrative that honors humanity's natural place in the world.

Comparing this educational goal to those repeatedly documented in common standardized assessments, it is difficult to argue where educators' time would be better spent. As an educational resource, *The Unsettling of America* is a testament to our historical roots in the land, a critique of our present abuse of it, and a call to action for our future relationship with it. For educators interested in promoting ecological consciousness, its content effectively informs instructional objectives in history, literature, government, or psychology courses. As a theme for developing interdisciplinary studies that purposefully connect schools with their larger community and learners to the natural world, lessons built around food issues can enhance this informative text.

Perhaps one of the most well-known contemporary initiatives toward community-based food education is Alice Waters' Edible Schoolyard in Berkeley, California. In her story of its inspiration, development, and successes, *Edible Schoolyard: A Universal Idea*, Waters recalls her first impression of Martin Luther King Jr. Middle School to a local journalist: "You want to see a great example of how *not* to use land? You should come look at this enormous school in my neighborhood that looks like nobody cares about it. Everything wrong with our world is bound up in that place and in the way we treat children."[16] In her critique, this revolutionary restauranteur and former teacher pinpoints an unfortunate truth about education in this country: we may tout its importance, but one need only look at the state of so many schools across the nation to recognize hypocrisy. The state of our schools—from physical structure to educational philosophy—is a local community issue as much as it is a governmental one; because "nobody cares," all suffer in what is "wrong with our world." Describing efforts to change this, Waters recounts how school leaders, teachers, students, and community volunteers helped create a productive one acre school garden and food education program. Perhaps most fascinating is that this book opens with photographs of children working in school gardens in the early 1900s. As Waters details the process of implementing food education and garden responsibilities in this middle school, there seems to be an implied question as to how and why we went away from gardening in schools in the first place:

> . . . there's just something infectious about the idea of an Edible Schoolyard—it's such a hopeful and uplifting notion that it just takes hold of people in the same way it took hold of me. It makes people want to help out, as a way of believing in a world where such a thing could really happen. I saw this spell

> fall over more and more teachers, as time went by, and also over volunteers who wandered in from their various lives, offering to help. . . .[17]

Waters' school garden with accompanying culinary program brought people together in a shared effort to improve school grounds, support local farms and businesses, educate about nutrition, and foster community compassion through eating locally harvested, self-prepared food. As Waters concludes in her appeal for large-scale transformation of school education, "these are the bedrock pleasures and values on which our agrarian democracy was founded, and they remain the key link between our private selves, the health of our bodies and our planet, and each American's power to influence positive change."[18] As with Berry's sentiments, who couldn't see the relevance of better understanding the health, social and environmental issues surrounding food? Isn't it worth the effort if even some personally engage with this knowledge?

In her essay "Fast-Food Values and Slow Food Values" Waters elaborates on why food as part of restructured educational reform makes sense:

> Our system of public education operates in the same strange, no-context zone of hollow fast-food values. In school cafeterias, students learn how little we care about the way they nourish themselves—we've sold them to the lowest bidder. . . . Children need only compare the slickness of the nearest mall to the condition of their school and the quality of its library and its cafeteria to learn that our culture considers them more important as consumers than as students.[19]

That youth are valued more as consumers than as students is evidenced by the endless fight for resources in schools and adequate pay for this country's educators. In a call to action, Waters asserts that "the public school system is our last best hope for teaching real democratic values that can withstand the insidious voices of those who would have us believe that life is all about personal fulfillment and personal consumption."[20] There is hope in education that teaches students to value each other more than materialism, to value time spent outdoors more than on obtaining the latest technologies. A focused school gardening and lunch program is one way to start because of its collective, sustained effort; this type of initiative invites crucial participants from the community—including parents, farmers, landscape artists, chefs, local businesspeople and architects—into the educational process. As many schools across the country have already discovered, school gardening and culinary projects also bring together administrators, teachers, students, and others such as cafeteria and custodial staff in meaningfully constructive ways. Imagine the collaborations and outcomes that emerge from sharing meals planned and prepared by students themselves; consider the pride students might take in designing and maintaining more intimate eating areas for themselves; think about how important composting and recycling could become for students if they learned the impact of their own wasted food or packaging. In terms of curricular needs, including food as part of education reform presents exciting, relevant opportunities for all subject areas, particularly sciences, health, math, business, economics, and design. As curriculum support, the humanities represent another level for food awareness, inviting related lessons in history, languages, literature, sociology, and philosophy. This endeavor not only brings students together but facilitates much needed cooperation among educators and administrators as well. All involved would be working together to empower students who understand how they can personally affect change.

Farmer and educator Michael Ableman, an advocate for food-based school reform, ascribes another intriguing metaphor to the state of public education and agriculture in this country: "Schools and farms have become a lot alike. They have both become factories, with assembly-line controls and engineered inputs, cranking out either grades and test scores or 'food.'"[21] Any administrator, parent, teacher, or student would be hard-pressed to disagree with this as we become further entrenched in standardized testing requirements and data-driven results. In his essay "Raising Whole Children Is Like Raising Good Food: Beyond Factory Farming and Factory Schooling," Ableman describes the detriment we have caused our culture by the "industrialization" of our food system and schools. Like Waters, his important activism supports the reunion of people with the earth, and offers this correlation between educational and cultural reform: "what we have really lost in both our food system and our education system is context, a sense of how things relate to each other."[22] Building on this with clear connections to Berry's analyses, Ableman notes that offering food-based reform could support understanding "local, biological, interpersonal, ecological"[23] relationships. For those involved, benefits include "knowing the people who grew the food, knowing that their families were paid a living wage, knowing that the land has been well cared for and protected from development, knowing that the food has not been assaulted with an array of chemistry and that its genetic makeup hasn't been messed with."[24] These interconnected relationships are relevant in all subject areas as an amazingly focused, purposeful lens through which to plan instructional strategies and goals. If educators can foster students' awareness about these important contexts, the positive impact their perspectives could have for human and environmental health could be exponential—for fair trade and local businesses; for conscientious, sustainable products and processes; for individual, familial, and cultural values. Let us face consumer culture by educating young people to act responsibly within it.

As Ableman astutely cautions, we cannot afford not to pursue these reforms. The seriousness of what is at stake may be only measured in full after it is too late:

> As we bear witness to the disappearance of nature, and the disconnection of our society from it, we also see an increase in confusion, an extreme lack of compassion and understanding of how to care for each other and for our world, a loss of understanding with regard to cause and effect.[25]

This "loss of understanding with regard to cause and effect" creates two very obvious conditions in our schools, which invariably extend to communities and to society as a whole. First, it means a potential disconnect between recognizing what we put in our bodies as beneficial or harmful to our health. Children whose families do not or cannot embody nutritious eating habits at home may never learn to appreciate the connection between wholesome food and feeling good. Second, the ways we regard (or disregard) natural resources and animals that contribute to our food sources may directly influence how we perceive and treat each other. Without the experience of having been around or been responsible for the care of vegetation or animals raised for food, perhaps it is more difficult to value how they are used or to be offended when they are abused. As Donna Harraway remarks in *When Species Meet*, "outside Eden, eating means also killing, directly or indirectly, and killing well is an obligation akin to eating well. This applies to a vegan as much as to a human carnivore. The devil is, as usual, in the details."[26] Developing relevant educational curricula to address details of health and social impacts (i.e., childhood diabetes and obesity; school violence and cyber-bullying) cascading from this disconnection will help us begin the journey toward healing with a more ecologically literate student body. It could mean that every individual gains knowledgeable recognition of how each of their choices matter. With this foundation, exercising free will may look extraordinarily different.

Fortunately, general public awareness about healthful food and agricultural practices is growing but we cannot afford to wait for its results to take hold. Perhaps adults learn to care more about food as we consider our grocery bills, health issues, and energy levels, as we plan meals and develop our palates, but by this time it may be difficult to change long held habits. Hopefully we learn early enough to pay attention to what we put into our bodies. Most of us will not. Given that "from a public health perspective, obesity is the most serious nutrition problem among children as well as adults in the United States,"[27] it is imperative to begin food education as early as possible. As professor and writer Marion Nestle suggests about the consumer culture at blame for this issue,

> Adults may be fair game for marketers, but children are not. Children cannot distinguish sales pitches from information unless taught to do so. Food companies spend at least $10 billion annually enticing children to desire food brands and to pester parents to buy them. The result: American children consume more than one-third of their daily calories from soft drinks, sweets, salty snacks and fast food. Worse, food marketing subverts parental authority by making children believe they are supposed to be eating such foods and they—not their parents—know what is best for them.[28]

If consumers cannot control the marketing that targets children, as Nestle suggests, then at least we could combat the problem through food education that starts at a young age and continues meaningfully throughout each child's education. A school in New York City is doing just that in its cafeteria and finding positive results based on student feedback. Juliet Pennington, a writer for the magazine *Vegetarian Times* reports that Public School 244, located in Flushing, Queens, has adopted an all vegetarian cafeteria menu for its elementary students based on input from students and parents in the district. A nutritionist interviewed—Maria Ankrom, MS, RD—supports the decision, noting that "'When children get exposed to healthy foods early in life, they are a lot more likely to carry these habits into adulthood.'"[29] The change has prompted positive discussions among parents, students and the school. Kids are talking about the health and ethics of food. Given the experience, they are appreciating the value of wholesome food. Speaking about advocating for a similar vegetarian menu in her next school, third grader Mariam Satti exclaims: "'I'm going to tell them that they should serve good food like they do here and not serve pink-slime chicken nuggets or anything like that . . . I hope they listen, but if they don't, I won't stop telling them until they do listen.'"[30] Mariam's determination reveals that just as young children can demand foods we know are bad for their health, they are also quite capable of requesting what is good for them. Adults might follow her attitude in working with education policy makers. Professors Troy Duster and Elizabeth Ransom suggest that "with less than 2 percent of the US population engaged with agriculture, a whole generation of people has lost valuable knowledge that comes from growing, preserving, and preparing one's own food"; they argue that the only way to "alter our eating habits so that we make a notable dent in the coming health crisis around the pandemic of childhood obesity and Type II diabetes" is through "long-term planning that will include going into the schools to change the way we learn about food."[31] If education can influence the way young people understand food by encouraging more healthful choices as five-year-olds, then it might positively change their long-term eating habits. And if they act on those conscientious habits as adults—consumers and voters—then we have a shot at exposing the food system for what it is and creating positive change.

Progress in cafeterias starts in classrooms. Eric Schlosser's popular *Fast Food Nation: The Dark Side of the All-American Meal* encourages better lunch habits along with, at the very least, thoughtful discussion about those habits in middle and high schools. To the question "what single thing could change the US food system, practically overnight?" he suggests education is key:

> Widespread public awareness—of how this system operates and whom it benefits, how it harms consumers, how it mistreats animals and pollutes the land, how it corrupts public officials and intimidates the press, and most of all, how its power ultimately depends on a series of cheerful and ingenious lies.[32]

His work advocates that we reestablish young people's understanding of cause and effect as it relates to the food system. If, as Harraway says, the "devil is . . . in the details" then the heart of Schlosser's message is to bring those details to the surface. Michael Pollan, author of *The Omnivore's Dilemma: A Natural History of Four Meals*, agrees that "nothing could do more to reform America's food system—and by doing so improve the condition of America's environment and health—if the rest of us were suddenly to weigh in."[33] Educating young people about their food choices and the impact of those choices on health, environment, and social welfare, teaches them how they can weigh in. In *Omnivore's Dilemma*, Pollan explains the danger of Americans' current disconnect with the food system, particularly the ingredients and processes that render food unrecognizable. He frames our overwhelming smorgasbord as akin to the danger hunter-gatherers faced in tasting unfamiliar finds such as poisonous mushrooms: "One way to think about America's national eating disorder is as the return, with an almost atavistic vengeance, of the omnivore's dilemma. The cornucopia of the American supermarket has thrown us back on a bewildering food landscape where we once again have to worry that some of those tasty-looking morsels might kill us. (Perhaps not as quickly as a poisonous mushroom, but just as surely.)"[34] In the classroom, teachers could incorporate Pollan's analogy in learning about any number of related issues, but the current political debates surrounding GMO food labeling especially come to mind as one that might encourage students to look more carefully at what they put in their bodies.

On the topic of GMOs, which are now common in vegetable and grain markets, we might return to Atwood's *Oryx and Crake* to consider genetic modifications of the animals we raise for food, as well as the conditions in which they are raised. In addition to the enormous, fearsome "pigoons" capable of harvesting human organs, the "ChickieNobs" lab-raised meat that have no heads and Jimmy's pet rakunk (a hybrid of a skunk and raccoon), Atwood more realistically portrays the impact of raising unnaturally oversized animals in conditions of crowded confinement that resemble the industrial food system's Concentrated Animal Feeding Operations (CAFOs). Specifically, Snowman's (Jimmy's) early childhood reflections on the plight of food animals ('why were the cows and sheep on fire?'[35]) offers critique on current practices of raising more meat for human consumption than sustainable practices allow. Alluding to the many meat recalls and contaminations this country and others have experienced, his father explains that disease had threatened the animals' and thus human health so they must be destroyed. As Jimmy watches the bonfire of bodies, readers see a boy conflicted. In a poignant metaphor about the ducks on Jimmy's rubber boots (whom he fears harming by walking through disinfectant) readers may pick up on another chilling commentary about our treatment of such animals: "He'd been told the ducks were only like pictures, they weren't real and had no feelings, but he didn't quite believe it."[36] Considering ongoing debates regarding the sentience of animals we raise for food or use in medical experiments, these "pictures" perhaps refer to supermarket images of animals like those on the bonfire—clean cellophane packaging that denotes old-fashioned pristine farms with happily grazing animals.[37] Throughout the trilogy are many engaging passages appropriate for introducing topics related to animal agriculture, genetic engineering, or marine life and the impacts of unsustainable fishing practices. As part of analyses that could come from these novels, a quite interesting literary distinction to make with our students is that which Atwood makes herself about science fiction and speculative fiction (which is how she classifies the trilogy): . . . "[Science fiction] denotes books with things in them we can't yet do or begin to do, talking beings we can never meet and places we can't go" whereas "[speculative fiction] employs the means already more or less to hand, and takes place on Planet Earth."[38] That the means for advancements Atwood presents in the trilogy are already in place ought to give us all pause.

A crucial advocate for animal welfare, who has also enhanced public awareness of autism, is Temple Grandin. Her lifetime work in animal science has profoundly improved the industry and can inform classroom discussions about animals raised for food, particularly in light of ethical implications that Atwood and others explore. In *Animals in Translation: Using the Mysteries of Autism to Decode Animal Behavior,* Grandin brilliantly explicates animal sentience and behavior based on her experience with autism, her science background, and her extensive familiarity with farm feed lots and slaughterhouses. On changing the food system she, like Schlosser and Pollan, also advises that we uncover the details: "My last recommendation is that farms and slaughter plants should have glass walls. I tell executives, 'There's this wonderful technology you can use to improve animal welfare. It's called glass. It's called webcam.' People need to see what's happening on farms and inside plants."[39] Grandin's advocacy of a more humane, symbiotic relationship with animals we raise for food does not suggest we adopt vegetarian diets. Instead, she urges that we support ways to raise animals for meat that honors the emotional behaviors science proves they have. This includes making relevant adaptations to living conditions that will stimulate their natural tendencies (engaging the curiosity and free movement of pigs, who we know are highly intelligent, for example) as well as ensuring humane practices in routine handling (such as Grandin's design for curved, tunnel-like chutes and corrals that help manage cattle's sensitivities to distractions). An excellent educational resource for introducing or exploring Grandin's work and ideas is Mick Jackson's 2010 film about her life, *Temple Grandin*. For more advanced research, in addition to *Animals in Translation*, interested students could consult Grandin's *Animals Make Us Human: Creating the Best Life for Animals* and *Humane Livestock Handling: Understanding livestock behavior and building facilities for healthier animals.*

Support for this kind of reform in animal husbandry as well for general agricultural practices also starts with farmers who recognize the inherent threat of current practices to our personal health, the condition of the environment, and the welfare of animals. Unfortunately, many of us do not have much opportunity to hear from them unless we are fortunate enough to frequent a local farm. Joel Salatin, owner of Polyface, Incorporated's "Farm of Many Faces" in Swoope, Virginia, is a farmer who writes and speaks about how we can improve our current systems. His books are a wonderful, accessible resource for anyone interested in learning more about these topics. For example, he introduces *The Sheer Ecstasy of Being a Lunatic Farmer* with a clear reference to the differences between family-style farms and agribusiness operations:

> As the industrial food system lashes back with innuendo and pseudo-science against the ecologically based food system, I think it behooves all of us to examine the differences between these two camps. People wonder how I can be such a happy farmer. The stereo-typical, unhappy farmer unfortunately is true much of the time. I hope this book will put in clear detail the depth and breadth of the difference between the chemical/industrial/global approach and the local/biological/ecological approach.[40]

Throughout, Salatin promotes his farming model which thrives on ecologically-friendly interactions between plant and animal species to restore soil devastated by improper farming practices. He also details how possible and profitable it is to raise happier, healthier animals who do not require extensive regimes of antibiotics or abusive confinements, as in many Concentrated Animal Feeding Operations (CAFOs). As Salatin relates in a chapter called "Relationships," there is a fundamental reason why we should care about the details of food we consume: "When people know their farmer, they connect viscerally with what is before them on the plate. After all, dining is a fairly intimate experience. Next to the act of marriage, eating is one of the more intimate things we do as humans. We take in this food, right into our bodies, and it becomes us. Flesh. Blood. Being. Mind."[41] This seemingly spiritual connection between food and our well-being is at the heart of Salatin's work. His passion for it is contagious. Bringing attention to the importance of local farms in a community as well as their economic, social and health benefits versus large industrial operations, Salatin explores the unsustainable destructiveness of dominant practices that exist here and in other developed countries.[42] Pairing Salatin's perspective with the following excerpt from David Brin's *Earth* invites students to investigate data related to the long-term consequences of these practices:

> Still, the 'great-die-back' everyone talks about always seems to lie a decade or so in the future, perpetually deferred. Innovations like self-fertilizing rice and super-mantises help us scrap by each near-catastrophe just in the nick of time. Likewise, due to changing lifestyles, few today can bear the thought of eating the flesh of a fellow mammal. Putting moral or health reasons aside, this shift in habits has freed millions of tons of grain, which went into inefficient production of red meat.[43]

Brin alludes to a forced cultural shift based on current habits of meat consumption worldwide. As meat continues to be available at cheaper prices, demand rises in a system that ignores looming shortages by consuming significant amounts of grain, water, and land. The context of this internationally relevant commentary offers plentiful interdisciplinary connections to world food production and distribution systems. An economics course, for example, might correlate a country's economic growth with meat consumption, while students assess potential benefits and consequences. Chemistry or agriculture connections might extend to learning about the importance of maintaining the world's seed banks or about the potential implications of seed monopolies like Monsanto. Health and pollution impacts of genetically-engineered crops and animals offer material for earth science and biology courses, while more complex specifics of innovation might be evaluated in higher level genetics or biochemistry classes. Further, a course in logic, marketing or journalism could follow the trends of advertisers' manipulation of data as new products or systems are marketed as panaceas. Though not the primary focus of the text, Brin's passage nevertheless offers much fodder for discussion and debate, particularly from an international perspective of developed and developing countries that demand increasing amounts of animal protein for human consumption.

Returning to Shelley's *Frankenstein* gives us another character conflicted about humanity's consumption of meat. In pleading with the scientist to grant him a companion, Frankenstein's creation invokes a more philosophical argument: "My food is not that of man; I do not destroy the lamb and the kid to glut my appetite; acorns and berries afford me sufficient nourishment. My companion will be of the same nature as myself, and will be content with the same fare."[44] Although merely a momentary critique in the text, Shelley's diction here—"destroy," "glut," and "sufficient"—open the door to a discussion of ethics, nutrition, and cultural practices regarding animals used for food. By engaging with philosophical, ethical, or different cultural arguments, students could also explore the creature's juxtaposition of human "nature" against that of his imagined "companion" who will be "content" without meat.

As these literary examples illustrate, food is one of the more sensitive issues that speculative, science, and climate fiction can help us broach. While in many ways food connects us to other cultures with the delectable array of flavorful experiences it provides, it can also divide us. In addition to socioeconomic inequities related to food and water (which will only increase with predicted changing weather impacts on land and growing seasons), food can be a lens for exploring diverse cultural beliefs. A contemporary piece that addresses how food influences our perception of cultural "Others" is the first of Octavia Butler's Patternist series—*Wild Seed.* In this story, readers meet the character of Anyanwu, a female demi-god, who is persuaded by a male spirit entity—Doro—to leave her people in pursuit of his. For the promise of producing children (with Doro) more like herself than had been the case the previous three hundred years with marriages to mortal men has allowed, she reluctantly leaves her home. On their journey toward his people's land, aboard what is essentially Doro's slave ship, the water (unknown territory for Anyanwu) initially frightens her. Aboard the ship she is isolated not only by race (she has encountered white people for the first time and they are slaveholders), but also by ancestry—most are Doro's people either by direct relation or pledged allegiance. Her initial impulse to jump overboard in a suicidal depression is redirected by her freedom as a shape-changer to transform into different human and other animal forms. Instead of death when she enters the sea, she traverses the waters as "the dolphin whose flesh she had eaten"[45] to learn how to become one. While she has sought Doro's permission to become a dolphin—for she loses her independence the moment she consented to journey with him—her immersion in the water marks at least a symbolic, temporary reclamation of freedom. In her new body, Anyanwu discovers that "she could close her eyes and perceive an only slightly diminished world around her with her ears. She could make sounds and they would come back to her as echoes bearing with them the story of all that lay before her."[46] Swimming with dolphins, she is surprised that "their chatter sounds more human now."[47] Fascinated by their behaviors and interactions, Anyanwu finds it "harder to think of [them] as animals" because "swimming with them was like being with another people. A friendly people. No slavers with brands and chains here."[48] No longer perceiving them as so different from herself, she loses her taste for their flesh. In addition to the richness offered in her personal conflict, this metaphorical passage might prompt important discourse of humanity's colonizing histories or of the sentience of higher level mammals like the dolphins Anyanwu experiences.

Throughout the novel, Anyanwu struggles in various ways to feel comfortable in Doro's world and "civilization." A food-related cultural conflict ensues when she discovers the cause of her stomach upset after enjoying a meal with Doro's people in New York: "It was milk that had sickened her. Animal milk! These people cooked many things with animal milk! She covered her mouth with her hand. Did Doro know? But of course he did. How could he not? These were his people!"[49] Anywanu is angry at her unwitting participation in what she considers a repulsive "abomination."[50] Animal milk, she insists to Doro, is for animals and "she had never in her long life lived among people who violated this prohibition."[51] She refuses to consume foods cooked with milk again although she reluctantly agrees to obey Doro's protestation to "let others have theirs in peace."[52] The cultural beliefs of students in the learning environment may determine how instructors explore this section of the text, but it provides a valuable opportunity to consider perspectives different from one's own.

Tom Standage's *An Edible History of Humanity* provides more complex opportunities for analyzing the cultural significance of food. By weaving an intriguing narrative of humanity's relationship with food into cultural histories, Standage emphasizes the vast influence food has in our lives. He describes his approach in the book's introduction, "Ingredients of the Past":

> There are many ways to look at the past: as a list of important dates, a conveyor belt of kings and queens, a series of rising and falling empires, or a narrative of political, philosophical, or technological progress. This book looks at history in another way entirely: as a series of transformations caused, enabled, or influenced by food. . . . [Food] has acted as a catalyst of social transformation, societal organization, geopolitical competition, industrial development, military conflict, and economic expansion. From prehistory to the present, the stories of these transformations form a narrative that encompasses the whole of human history.[53]

A text such as this paired with any number of literary, philosophical, economic, or scientific texts could provide a more comprehensive understanding of not only our country's food history, but of the world's. Standage's use of food as a centralizing theme connects more abstract, complex concepts to an easily appreciated common motivator for all societies. His references can support students' understanding of their own values by providing insightful historical contexts for understanding how those values evolved in different cultures. This perspective can then inform how they think about as well as react to current food-related crises around the world. Students studying famines throughout the African continent, for example, could analyze the political, historical, environmental and social circumstances that contribute to forced migrations and widespread starvation. Along with obvious interdisciplinary connections, a discussion about human population, an analysis of global food distribution, or an evaluation of climate-related food and water predictions could all be better informed by a blend of historical, current, and future considerations.

We need to start somewhere and educators have much to contribute. Instruction that places food-based environmental issues at its core will prepare students for the real problems we face as a global community. Explaining the importance of comprehensive reform in her essay "Tapping the Well of Urban Youth Activism: Literacy for Environmental Justice," Literacy for Environmental Justice (LEJ) founder Dana Lanza reinforces why ecopedagogy must be cross-curricular:

> Systemic change to long-standing social constructs usually requires proponents who are literate in science, civics, and history, and proficient in writing and speaking. . . . [Students] must learn to leverage strong communications skills with compassion, honesty, and conviction. Perhaps most important, they must develop strategic planning skills and the patience to find the way to change one step at a time.[54]

Schools can work toward this change by implementing curriculum-supported environmental projects that foster collaborative food experiences like campus gardens or community meals. However, effective systemic change requires comprehensive programming that will support individual school efforts as part of broader education reform. For educators to provide quality instruction that fosters knowledgeable action within a global environmental lens, they deserve access to relevant professional training, development, and resources. These supports should be developed by stakeholders who can provide expertise as well as action. This team could include, though is not limited to, farmers, scientists, environmentalists, politicians, community and business leaders, human and animal rights activists, nutritionists, administrators, teachers, parents, students. Inviting contributions from all stakeholders puts education at the forefront of community goals, reinforcing the popular proverb "It takes a village. . . ." To Alice Waters' point, it could put schools, rather than malls, in a central, valued position.

Daniel T. Blumstein, scientist and professor at the Institute of the Environment and Sustainability, and Charles Saylan, executive director of the Ocean Conservation Society, also emphasize the importance of direct student engagement in these efforts. Discussing it in *The Failure of Environmental Education [And How We Can Fix It].*, they frame it as such:

> Consider that environmental awareness may not lead to action at all. Instead it may be some combination of personal involvement and emotional investment

> that prompts people to act, and environmental awareness may only help frame how they will act. If this is true, and action is the goal, educational efforts will have to bring students around to where they feel a personal commitment to a given project or topic.[55]

If teachers offer students opportunities to commit in such a way to projects involving research they have done on their local environment, consider the progress that might be made toward improving one's sense of "home," including the stewardship which comes with such an attachment. Ecological community projects (food-based or otherwise) certainly contribute to this goal. An example referenced by Saylan and Blumstein reveals how grassroots ideas with community improvement in mind can have long-term transformative effects for school communities. Crenshaw High School's attempt to heal their community after the 1992 Los Angeles riots included a "two-acre fruit and vegetable garden on an abandoned lot adjoining the school" which eventually became a company that "donated 25 percent of what the students grew to needy people in the community. The balance of fresh goods, along with salad dressings developed by students, were sold nationwide for profit, with half the proceeds going to the operation of the company and the other half allocated to scholarships for student program managers."[56] Importantly, the authors offer reassuring guidance for its extraordinary results, which might be a relief to educators wondering how to implement such inspiring yet complex projects:

> We believe one of the primary reasons that Food From the Hood was so successful was that it was open-ended. No one at its inception had any idea it would grow as it did. Students and involved teachers alike had to learn, on the fly, how to deal with the challenges of running a fast-growing business. Goals were adjusted as the project took on a momentum of its own. Food From the Hood became a remarkable example of team effort, rewarding participants with a level of practical confidence unlikely to occur through purely academic endeavors. . . . Seeing an idea through from concept to reality is a powerful experience. The process promotes creativity, resourcefulness, discipline, and perseverance. It requires the integration of knowledge across disciplines, as well as leadership and communication. It is a wonderful preparation for life.[57]

Indeed, and the learning experience far more meaningful than what can be objectively assessed. Teachers need not be experts in gardening or culinary arts or business to make a difference. Part of the joy in projects and curricula like this is in learning something new with your students.

In *Animal, Vegetable, Miracle: A Year of Food Life,* Barbara Kingsolver discusses how agricultural knowledge has essentially disappeared from our culture, adding that "we also have largely convinced ourselves it wasn't too important. Consider how Americans might respond to a proposal that agriculture was to become a mandatory subject in all schools, alongside reading and mathematics."[58]

I will leave this for the reader to consider but reiterate that there are many ways we can mobilize environmental consciousness around this "small, good thing" that is food. We need communities that believe in their schools and schools that position students as thoughtfully contributing members of their communities. An education paradigm shift toward ecological and social justice can help us reimagine the concept of citizenry at the global level so that we might act not merely as citizens of our particular place but as members of a global community, inhabitants of a shared planet.

The work ahead overwhelms, but every step taken will bring us closer to realization, action, and satisfaction that leads to even more steps forward. Changes in one classroom can lead to changes in the classroom down the hall to changes in the cafeteria, the outdoor space, the building structure, and so on. As Joel Salatin humorously says of "the sheer ecstasy of being lunatic farmers," who believe in a better way forward, "I hope you will catch the vision. If the lunatic fringe is where the truth is, I hope you will aspire to be a lunatic too. Let's enjoy it, embrace it, and have fun with it."[59]

Let's share the table with our students.

NOTES

1. David Orr, *Dangerous Years: Climate Change, the Long Emergency, and the Way Forward* (New Haven: Yale University Press, 2016).
2. Richard Beach, Jeff Share, Allen Webb, *Teaching Climate Change to Adolescents: Reading, Writing, and Making a Difference* (New York: Routledge, 2017), 132.
3. Wendell Berry, *The Unsettling of America* (San Francisco, CA: Sierra Club Books, 1986), dedication.
4. Ibid., 19.
5. Olmert, Meg Daley. *Made for Each Other: The Biology of the Human-Animal Bond.* (Da Capo Press, 2009), 198.
6. Ibid.
7. Rebecca Young, "McLovin: Our Food and What We Can Do About It," *Ragazine, The Global Online Magazine of Arts, Information, and Entertainment,* November-December, 2014, 10, no. 6.
8. Wendell Berry, *The Unsettling of America*, 56.
9. Carly Lettero, "Spray Glue Goes. Maggots Stay.," in *Moral Ground*, 104.
10. Ibid.
11. Ibid., 106.
12. Ibid., 107.
13. Lesson idea also shared in *Teaching Climate Change to Adolescents: Reading, Writing, and Making a Difference,* 36.
14. Wendell Berry, *The Unsettling of America,* 56.
15. Ibid., 58.
16. Alice Waters, *The Edible Schoolyard: A Universal Idea* (San Francisco, CA: Chronicle Books LLC, 2008),10.
17. Ibid., 30.
18. Ibid., 40.
19. Alice Waters, "Fast-Food Values and Slow Food Values," 54.

20. Ibid., 54.
21. Michael Abelman, "Raising Whole Children Is Like Raising Good Food: Beyond Factory Farming and Factory Schooling," in *Ecological Literacy: Educating Our Children for a Sustainable World*, ed. Michael K. Stone and Zenobia Barlow (San Francisco, CA: Sierra Club Books, 2005), 178.
22. Ibid., 179.
23. Ibid.
24. Ibid.
25. Ibid., 183.
26. Harraway, Donna J. *When Species Meet*. (Minneapolis: University of Minnesota Press, 2008), 296.
27. Marion Nestle, "One Thing to Do about Food," ed. Alice Waters, *The Nation,* September 11, 2006, 14.
28. Ibid.
29. Juliet Pennington, "The Cafeteria Report," ed. Elizabeth Turner, *Vegetarian Times*, September 2013, 56.
30. Ibid.
31. Troy Duster and Elizabeth Ransom, "One Thing to Do about Food," ed. Alice Waters, *The Nation,* September 11, 2006, 17.
32. Eric Schlosser, "One Thing to Do about Food," ed. Alice Waters, *The Nation,* September 11, 2006, 14.
33. Michael Pollan, "One Thing to Do about Food," ed. Alice Waters, *The Nation,* September 11, 2006, 16.
34. Pollan, Michael. *The Omnivore's Dilemma: A Natural History of Four Meals* (New York: Penguin Books, 2006.).
35. Margaret Atwood, *Oryx and Crake* (New York: Random House, 2003), 19.
36. Ibid., 15.
37. See also Rebecca Young, "McLovin: Our Food and What We Can Do About It," *Ragazine, The Global Online Magazine of Arts, Information, and Entertainment,* November-December, 2014, 10, no. 6.
38. Margaret Atwood, *The Handmaid's Tale and Oryx and Crake "In Context.," PMLA* vol. 119, no. 3, Special Topic: Science Fiction and Literary Studies: The Next Millennium (May 2004), pp. 513–517, 513.
39. Grandin, Temple and Johnson, Catherine. *Animals In Translation: Using the Mysteries of Autism to Decode Animal Behavior*. New York: Scribner, 2005), 228.
40. Salatin, Joel. *The Sheer Ecstasy of Being a Lunatic Farmer*. (White River Junction, Vermont: Chelsea, 2010), xvi.
41. Ibid., 253.
42. See also Rebecca Young, "McLovin: Our Food and What We Can Do About It," *Ragazine, The Global Online Magazine of Arts, Information, and Entertainment,* November-December, 2014, 10, no. 6.
43. David Brin, 43.
44. Mary Shelley, *Frankenstein with Related Readings* (New York: Glencoe McGraw-Hill, 2000), 107.
45. Octavia Butler, *Wild Seed* in *Seed to Harvest* (New York: Grand Central Publishing, 2007), 78.
46. Ibid.
47. Ibid., 79.
48. Ibid.
49. Ibid., 109.
50. Ibid., 110.
51. Ibid., 111.
52. Ibid.
53. Tom Standage, *An Edible History of Humanity* (New York: Walker Publishing Company, Inc., 2009), ix.

54. Dana Lanza, “Tapping the Well of Urban Youth Activism: Literacy for Environmental Justice,” in *Ecological Literary*, 218.
55. Charles Saylan and Daniel T. Blumstein, *The Failure of Environmental Education (And How We Can Fix It)* (Los Angeles: University of California Press, 2011), 109.
56. Ibid., 110.
57. Ibid., 110-111.
58. Kingsolver, Barbara. *Animal, Vegetable, Miracle: A Year of Food Life*. New York: HarperCollins Publishers, 2007), 9.
59. Salatin, Joel. *The Sheer Ecstasy of Being a Lunatic Farmer*, xvi–xvii.

Chapter Five

Morality and Environmental Responsibility

An Interdisciplinary Reading of Franzen's Freedom

Sharing the table with our students, with one's children, and with each other opens the opportunity for honest conversations about what it means to live responsibly and respond proactively to the climate crises we are confronting. A powerful lens that can reflect, critique, and inform individual and social behavior, fiction nurtures our capability to navigate this changing world. In *The Great Derangement: Climate Change and the Unthinkable*, Amitav Ghosh suggests that

> . . . what fiction—and by this I mean not only the novel but also epic and myth—makes possible is to approach the world in a subjunctive mode, to conceive of it *as if* it were other than it is: in short, the great, irreplaceable potentiality of fiction is that it makes possible the imagining of possibilities. And to imagine other forms of human existence is exactly the challenge that is posed by the climate crisis: for if there is any one thing that global warming has made perfectly clear it is that to think about the world only as it is amounts to a formula for collective suicide. We need, rather, to envision what it might be.[1]

While Ghosh laments that "this challenge has appeared before us at the very moment when the form of imagining that is best suited to answering it—fiction—has turned in a radically different direction," the instructive quality of witness literature that has emerged can engender a paradigm shift toward environmental and social justice education. Pairing nonfiction with fiction supports interdisciplinary goals toward this endeavor and fosters greater depth for analysis.

While previous chapters focus on a range of works that readily lend themselves to inclusion in American public education curricula, this chapter presents a more in-depth model of such an approach appropriate for higher level literature or science classes, pairing Jonathan Franzen's cultural fiction *Freedom* with excerpts from Patricia S. Churchland's theoretical text *Braintrust: What Neuroscience Tells Us about Morality*. The analysis is an adaptation of an assignment I found both challenging and rewarding in SUNY Binghamton professor Leslie Heywood's course Literature and/on the Brain, which asked students to analyze a fiction text from the lens of one of the nonfiction selections we studied. Appropriate for higher level readers in a literature, science, sociology, or psychology course, Franzen's *Freedom* and Churchland's *Braintrust* (along with other similar literature/science pairings), help explicate through narrative the science of human moral behavior toward each other and the environment. Relationship dynamics between selfish and altruistic characters in Franzen's *Freedom* expose moral and ethical implications for human behavior in current culture; examining the link between neuroscientific-based motivations and socially productive expectations, Franzen's novel promotes healthy reconciliation of human beings' competitive and prosocial selves in order to "manage well in the existing social ecology."[2] The analysis that follows offers a scientific lens for studying fiction that encourages empathic response to motivate individual and social action. Considered from this perspective, the aptly named *Freedom* offers opportunities to explore the Taker-Leaver decisions we make in our daily lives—whether we are aware of them or not—by engaging with characters who confront them on multiple levels.

Professor of Philosophy Patricia S. Churchland introduces *Braintrust* with the hypothesis that "what we humans call ethics or morality is a four-dimensional scheme for social behavior that is shaped by interlocking brain processes."[3] She explores these processes, offering neuroscientific data throughout her commentary on individual and social mammalian behavior as it applies to evolutionary, social and cultural constructions or implications regarding morality. Churchland notes that informing our individual and social decision making are "values rooted in the circuitry for caring—for well-being of self, offspring, mates, kin, and others."[4] Her research synthesizing evolutionary brain processes and recent discoveries in neuroscience illustrates a four-part framework for understanding what motivates human perceptions of and actions for healthy social living. In brief, the processes are

> (1) *Caring* (rooted in attachment to kin and kith and care for their well-being), (2) *recognition of others' psychological states* (rooted in the benefits of predicting the behavior of others), (3) *problem-solving in a social context* (e.g.,

> how we should distribute scarce goods, settle land disputes; how we should punish the miscreants), and (4) *learning social practices* (by positive and negative reinforcement, by imitation, by trial and error, by various kinds of conditioning, and by analogy).[5]

As highly social animals, humans are driven biologically to form close attachments with others. Throughout history, these bonds have enabled humans to benefit from group socialization contexts (from sharing basic survival strategies to developing complex support systems). Applying Churchland's attachment framework to current social and environmental dilemmas, however, reveals serious implications.

As Churchland explains in a chapter called "Caring and Caring For," a human being's primary concern is for self-preservation: "All nervous systems are organized to take care of the basic survival of the body they are part of. From an evolutionary perspective, the general point is straightforward: self-caring is selected over self-neglect."[6] The processes by which humans extend care to others are more complex but "the core idea is that in mammals, evolutionary adjustments in the emotional, endocrine, stress, and reward/punishment systems effectively extend the range of individuals whose well-being the animal cares about, at least for a certain set of survival-relevant behaviors."[7] According to Churchland, values extending from care systems apply first to ourselves, then our close-kin, and finally, our "affiliates."[8] A humanities-focused perspective on this topic encourages the examination of how science informs individual and social behaviors and, by extension, how these behaviors inform the sciences.

In *Freedom*, Franzen illustrates individual and social behaviors via the Berglunds—an American family whose fall from grace leaves them seeking redemption for themselves, each other, and the social culture to which they belong. What readers find in their characterizations and conflicts is a commentary on individual, familial, and social responsibility. This family lens illuminates on a larger scale the ethical implications of personal choices on cultural dilemmas: some readers will be intrigued by the Berglunds, and thus care about the moral questions that haunt and motivate them. Others will care about the social and environmental issues Franzen explores in the text. Either way, the challenge is to understand what determines the moral choices and behaviors that impact society. Authors provide the tool, readers wield it.

For students, it is helpful to consider reviewer Paul Charles Griffin's assessment of the objectives *Freedom* achieves. In "Sound and Fury, Signifying . . . What, Exactly?" Griffin offers an intriguing perspective of how audiences engage with this story: "the real questions that Franzen and *Freedom* explore are socio-aesthetic and eco-political. First, what is the role of the novel in contemporary society? Second, what the hell are we going to do, politically speaking, about the environmental crisis?"[9] And while some reviewers denounce Franzen's efforts to facilitate this conversation through his narratives or to propel his stories toward solution and change, others find achievement in Franzen's stated mission as a novelist: "to try to provide a bridge from the inside of me to the inside of somebody else."[10] As Griffin asserts, Franzen's "mission is spiritual in nature: to alleviate the lonely suffering of the individual in an increasingly atomized world, to create community through ritual introspection."[11] A reading of *Freedom* through the interpretive lens provided by Patricia Churchland's work on neuroscience and morality provides this "ritual introspection," which may well enhance personal interest and individual ability to "create community."[12] This transformative experience is exactly the reason literature is so essential to individual and cultural growth. *Freedom* initiates critical thinking and reflective action demanded by some of the most pressing global issues: human overpopulation, species extinction, environmental degradation, political and corporate corruption.

Asserting that the crux of the novel and these issues resides in a particularly American view of its title concept as linked with our "secularizing impulses of 'power,'" writer Sam Tanenhaus aptly names his review of *Freedom* "Peace and War": ". . . As each of us seeks to assert his 'personal liberties'—a phrase Franzen uses with full command of its ideological implications—we helplessly collide with others in equal pursuit of their sacred freedoms, which, more often than not, seem to threaten our own."[13] Franzen explores this clash through the individual, familial and social conflicts faced by an archetypal American family. Importantly, as Tanenhaus points out, "The Berglunds, introduced as caricatures, gradually assume the gravity of fully formed people, not 'rounded characters,' in the awful phrase, but misshapen and lopsided, like actual humans."[14] Like readers, members of the Berglund family struggle to balance their responsibilities and freedoms regarding individual desires, familial expectations, and social responsibilities. Admiring Franzen's development of this dynamic, novelist and critic Philip Hensher's "Family Fallout" offers its relevance in American reading culture:

> We don't choose our siblings, our parents, our children, but from day to day we choose, with the full volition of our existence, how those relationships will shape us. The interplay between our free will and the world which is wished upon us is the core, irreducible territory of great novels. Perhaps now that family ties are growing more dependent on volition and goodwill (or bad) rather than on duty and obligation, family is becoming more of a fruitful territory for the novelist's investigations, rather than less so.[15]

Franzen's intricate development of the Berglund family structure— their love and cruelty toward each other, their selfish expectations for and selfless forgiveness of one another—illustrates in practical form the duality of family and social life: it is hard to live together, and it is hard to live alone.

In this vein, each of Franzen's characters represents an unbalanced model of Patricia Churchland's neuroscientific explanation of how humans (and other social animals) form attachment bonds with others. For explication, my review will address the novel's three main characters—Walter, Patty, and Richard— including Richard (later referred to in the novel as Katz) as part of the Berglund family unit. Franzen explores characterizations and conflicts through Walter, Patty, and Richard that situate his text and their story in a uniquely American context—Walter struggles to balance his career philanthropy and inner goodness with a barren personal life and crumbling family relationships; in failed attempt to redeem her own childhood, Patty sacrifices her personal identity striving to be the perfect suburban housewife and mother; Richard repeats meaningless relationships with women as he fails to balance his growing musical fame with recovery from abandonment by his mother. The novel traces the interactions among these three characters from young adulthood to middle age simultaneously offering critique of their confining expectations on one another and praise for the emotional balance they create in each other. Franzen's development, dismantling, and ultimate reconciliation of their individual and collective friendships, Walter and Patty's marriage, as well as Patty and Richard's affair, illuminates with honesty and clarity the challenges we all face in expectations of ourselves and others.

Applying Patricia Churchland's work on neuroscience and morality to Franzen's character development of Walter, Patty, and Richard gleans a better appreciation for the evolutionary brain systems that motivate their attachments to one another and inform their individual and social morality (or lack of in the case of Patty and Richard). This is the impact of literature in general and Franzen's talent in particular. As Leo Robson comments in his essay "I Feel Your Pain: The Difficulty of Being Jonathan Franzen," Franzen's skill lies in his

> ability to place the reader right in the thick of his characters' lives, so that their pain becomes ours. He baldly states their problems and greedily ransacks their thoughts, and although he prefers rumination and paraphrase to the dramatic scene, his discoveries about a character's intentions or ideals are repeatedly tested in the sphere of action against their conduct.[16]

In other words, readers' literary discourse is informed by the characters who, like it or not, resemble us in thought and action. If readers engage empathically with a text more deeply than accepting an author's resolution of character conflicts, the benefit is introspection that respects the instructive nature of literature: the responsibility for change is not the task of the author but of the reader. Potential for individual and social action—in this case familial and environmental—results from how readers engage with the issues presented by the text once they have finished it.

Walter Berglund—in Churchland's attachment model—represents the extreme of prosocial behavior. While Churchland argues evolutionary nature is to care first for ourselves, then our close-kin, and then our selected affiliates, Walter is selfless and offers his full care and attention to his wife, Patty, and to social affiliates less fortunate than himself—particularly within the social dilemmas for which he seeks solution. As presented in the exposition of the novel, "Walter's most salient quality, besides his love of Patty, was his niceness. He was the sort of good listener who seemed to find everybody else more interesting and impressive than himself."[17] Unlike current cultural priorities that promote corporate greed and individual advancement, Walter "failed to thrive [as an attorney] and was shunted into outreach and philanthropy, a corporate cul-de-sac where niceness was an asset."[18] Eventually he expands his efforts by working for the Nature Conservancy. Our narrator's commentary on this move reveals its implication for Walter's familial obligations: "the only outward change in his life was his new scarcity at home on weekends."[19] At this point in the novel, readers see a Walter whose personal and family life suffer neglect because of his extreme prosocial community interests.

In his wife's case, we initially find that "for all queries, Patty was a resource, a sunny carrier of sociocultural pollen, an affable bee."[20] In Churchland's model, Patty neglects self-preservation, opting instead for attachment to close-kin and, superficially, her affiliates (in this case, her neighbors). The quintessence of motherhood and apparent neighborliness, "she might have been carrying all the hours of her day in the string bags that hung from her stroller."[21] This façade is quickly dismantled by her neighbor Merrie, however, to reveal "something rather hard and selfish and competitive and Reaganite in Patty; it was obvious that the only things that mattered to her were her children [Jessica and Joey] and her house—*not* her neighbors, *not* the poor, *not* her country, *not* her parents, not even her own husband."[22] An important addition to this—"Merrie's opinion"—is that self is also not important to Patty. While Walter embodies prosociality to the detriment of self and family (close-kin), Patty inhabits her role as parent much more completely—even to the exclusion of her independent fulfillment as a person or authentic participation in this community.

When readers first meet the third member of the Berglund's family unit, Richard—Walter's college roommate and friend and Patty's unfortunate crush—he is described by Eliza (Patty's drug addicted college friend and Richard's current lover) as "so big, it's like being rolled over by a neutron star. It's like being erased with a giant eraser."[23] His punk band, appropriately called the "Traumatics," and his side work in "demolition" suit his developing personality as a selfish manipulator of others—women in particular. As Franzen later develops, "Richard had no relationship with his mom"[24] which may, in Churchland's model, explain his inability to properly bond with others—he has no close-kin (until Walter) and his dismissal of affiliates (friends and fans) leaves him rather alone for most of the novel, except for vacuous affairs with various women. In short, at least initially, Richard is concerned about Richard.

Interestingly, Franzen's characterizations of Walter, Patty, and Richard illustrate three uniquely imbalanced care systems—and each essentially fails; Patty's autobiography *Mistakes Were Made* presents the conflicts that reveal why. As Patty relates about her and Walter's dismal sex life—partially a result of the "totality of her immersion in being Jessica and Joey's mother,"—"craving sex with her mate was one of the things (OK, the main thing) she'd given up in exchange for all the good things in their life together."[25] Essentially, readers learn that Patty marries Walter for his goodness and for his fiercely loyal belief in *her* goodness. She spends a large part of their married life trying to live up to this—and heal from her own childhood neglect and traumatic experience—through her devotion to their children (close-kin in Churchland's model). When Joey decides to leave home and move in with his girlfriend Connie Monaghan and her family, Patty's neighborliness is transformed as she becomes "a different kind of neighbor, a much more sarcastic neighbor."[26] "The move was a stunning act of sedition and a dagger to Patty's heart,"[27] indicated by her physical change: "Patty was looking a mess, gray-faced, poorly slept, underfed."[28] Their neighbor, Seth, aptly points out the cause: "Theft of her cub—the ultimate crime."[29] Since Patty's adult identity has been formed solely by her roles as mother and wife (but particularly mother), she has essentially denied her individual identity as a competitive athlete. As a result, unsuccessful attempts to reconcile with Joey lead her into isolation and depression. Ultimately, her grief and growing distance from her family, which initiated a renewed focus on herself, leads to an affair with Richard at the Berglunds' lake house.

This event eventually defines Richard's imbalance of care as well. Although their backgrounds and interests make them somewhat impractical friends, Franzen develops the important close-kin bond between Richard and Walter, whose moral goodness comforts Richard in an increasingly corrupt political world: "Yo, Jersey City here. Wondering if you can make me feel better about the situation in Kuwait. Give me a call."[30] As Patty astutely notes, she and Walter were Richard's "lifeline to a world of sanity and hope,"[31] their family becoming more important to him than his own. Early in the novel and despite an attempt "to put childish things behind him and sustain a real relationship with a grownup woman,"[32] Richard's selfish, independent "nature finally caught up with him"[33] to the extent of "finding himself homeless, at the age of forty-four, in midwinter, with maxed-out credit cards and a $300 monthly storage bill for all his crap."[34] With Walter's offer to work on the lake house for room and board, Richard finds himself in midst of Walter and Patty's conflicting opinions about Joey and their now unstable marriage—Walter has thrown himself further into work while Patty, self-pitying and spiteful at her perceived loss, has resorted to alcohol. What Patty sees of herself through "Richard's eyes" forces her out of depression and into action. She "got herself into shape: took Richard's advice about the drinking, started running and eating again, gained enough weight to fill in the most haggard of the lines that had been forming in her face, and generally acknowledged realities about her physical appearance which she'd been ignoring in her fantasy world."[35] The relationship they develop behind Walter's back serves both of their needs in the short term.

Patty pays attention to herself and her appearance, reconnecting with her previous athletic self (through old basketball friends) and her sexuality (toward Richard). Her priorities have shifted, importantly, from close-kin to self at the precise moment that Richard is attempting to "[get] his life sorted out"[36] by relying on the support of close-kin (the Berglunds) in the aftermath of his total self-absorption. Significantly, inconsistencies and imbalances in Walter's care system are also revealed by Joey's abandonment of his parents.

As evidenced in one of their early conflicts, Walter values prosociality over the individual gain characterizing competition and capitalism. This is demonstrated in Joey's first attempt to make money fraudulently: "Joey, making money is not a right. You're selling junk those girls don't really need and some of them probably can't even afford. That's why Connie's school has a dress code—to be fair to everybody."[37] A discussion of individual morality based on care systems becomes complex here because Walter Berglund appears to be the ideal citizen—environmentally conscious, politically proactive and selflessly philanthropic. Reading Walter's character through the lens of Churchland's work in morality, however, is telling. His motivations seem attributable to the childhood neglect he experiences while his older brother Mitch's purely self-motivated behaviors were rewarded. Mitch lived free of responsibility at home and "blew his paychecks on girls, drink, guns, fishing equipment, and his souped-up Thunderbird," while Walter performed motel maintenance duties to help alleviate the load of his overworked mother and support the family business. As an adult, Walter seeks solace and fulfillment in prosocial behavior that was not valued in his childhood. Since his care toward social others ('affiliates' in Churchland's terminology) is still not rewarded by his close-kin (Patty, Jessica, and Joey), Walter throws himself further and further into it, perhaps seeking acknowledgment in society's gratitude and the personal reward of serving others.

One of the aspects of human sociality Churchland develops in *Braintrust* is cooperation based on trust. In the chapter "Caring and Caring For," Churchland outlines and discusses the "crucial steps that lead from *only* self-caring, to the variety of kinds of sociality (*other-caring*) typical of mammals."[38] In "Cooperating and Trusting," she explores how "trusting, cooperative interactions can regularly occur between unrelated friends and strangers" within the established premises that "moral values are anchored by the neurobiology of sociability" and that "cooperation is an important morally relevant behavior."[39] If Walter has grown up to detest his brother's lack of cooperation with close-kin (leading to Walter's increased responsibilities to the family and the neglect of individual pursuits), it seems logical that he seeks cooperation with social affiliates when efforts with his wife and children also begin to fail. As Churchland notes, "Doing good feels good—at least sometimes,"[40] which is why Walter's character thrives by responding to social dilemmas in which cooperation and trust are perceived as valuable. It is thus ironic that while Walter is away for a Nature Conservancy fundraiser, Richard and Patty—arguably his two closest kin—turn to each other for their own healing in guilty betrayal of Walter. Patty gives in to the "ruthless and well-organized party of resistance fighters"[41] which cause her pulse to race, prompting important self-reflections, "Am I alive? Do I possess a body?"[42] That they succumb to their mutual physical attraction is disturbing in one sense and quite essential in another. As the narrator notes of Patty's grief afterward, "She'd taken advantage of a person who was messed up and susceptible but nevertheless trying hard to maintain some kind of moral order in his life. And so she was crying for Richard, too, but even more for Walter, and for her own unlucky, wrongdoing self."[43] Still, the affair (specifically its revelation to Walter) is ultimately the catalyst that heals, or balances, each of them: basic emotional impulses initiate the cognition that will redefine their characters.

Franzen's presentation of the affair's resulting consequences—Patty's further isolation and depression from guilt, Richard's aloofness toward Walter and persistent interest in Patty, and Walter's increasing role in philanthropic and political endeavors—illuminates the failings of each character's unbalanced care spheres. As Churchland points out about human morality, ". . . from a biological point of view, basic emotions are Mother Nature's way of orienting us to do what we prudentially ought. The social emotions are a way of getting us to do what we socially ought, and the reward/punishment system is a way of learning to use past experiences to improve our performance in both domains."[44] As each character moves through basic social emotions resulting from their behaviors, their reward and punishment system forces them to better balance their care spheres. Walter's relationship with Lalitha (his vibrant, exotic, ambitious intern) fosters his self-preservation (and reward system) while also encouraging care for close-kin through his potential interest of fathering a child with her. Eventually, his reward of caring for self (manifested in his sexual and love relationship with the adoring Lalitha) leads to his reconciliation with Joey. Walter's inadvertent dealings with the corrupt Cerulean Mountain Trust reestablish his close-kin attachment to his son in their shared predicaments. Later, Walter responds to

Joey's call about being "in some trouble" regarding the damaged military truck parts with "Hey, so am I! So is everybody!"[45]

Recognition of his own political compromise allows Walter to forgive and advise Joey, ultimately reuniting them as father and son. Both Richard and Patty essentially are forced to seek attachment bonds in social affiliates because of their distanced relationships with close-kin (namely Walter, but also Joey and Jessica in Patty's case). Richard's revelation of the affair to Walter (by sharing Patty's autobiography with him) represents, at least in his mind, the suppression of his own needs for those of another. He loves Patty but also loves Walter (and recognizes Patty's love for him) and believes he is protecting the marriage by walking away from her as he did in college. In the life he then builds for himself, readers see a musician who has accepted his fans and has learned how to use his music in more productive ways (i.e., participation in the band festival for the environmental group Free Space). Patty, after Walter finally leaves her, reinserts herself in her community (environment) by working in a local school, establishing friendships with coworkers and exerting care for other children (affiliates) in the absence of her own. Resolved in the conclusion of the novel, Richard's symbolic and Patty's actual repaired relationship with Walter completes the balance of care spheres for all three characters.

A closer look at the science of morality and Walter's characterization within this section of the novel helps educators answer students' interrogation, "*Why should we read this stuff?*" In fact, each character's returned balance of care for self, close-kin, and affiliates should elicit the type of introspection Griffin describes—a self-reflection that leads to community building. In the epilogue to V. S. Ramachandran's *The Tell-Tale Brain: A Neuroscientist's Quest for What Makes Us Human*, Ramachandran concludes that

> [s]elf-awareness is a trait that not only makes us human but also paradoxically makes us want to be more than merely human. As I said in my BBC Reith Lectures, 'Science tells us we are merely bodies of beasts, but we don't feel like that. We feel like angels trapped inside the bodies of beasts, forever craving transcendence.' . . . That's the essential human predicament in a nutshell.[46]

Walter seeks this transcendence for himself and for humanity in his work with the Cerulean Mountain Trust, and eventually, his and Lalitha's campaign against overpopulation—Free Space. Unfortunately for Walter—and for humanity—his altruism leads to the collapse of his close-kin relationships and activism burnout. After Patty leaves and moves in with Richard, Walter finds himself feeling—as might Franzen's reader—that

> [t]he world was moving ahead, the world was full of winners, LBI [the defense contracting company that both Walter and Joey are involved with] and Kenny Bartles cashing in, Connie going back to school, Joey doing the right thing, Patty living with a rock star, Lalitha fighting her good fight, Richard going back to his music, Richard getting great press for being far more offensive than Walter, Richard charming Connie, Richard bringing in The White Stripes. . . .while Walter was left behind with the dead and dying and forgotten, the endangered species of the world, the nonadaptive. . . . [47]

This sense of hopelessness may echo readers' own feelings toward environmental or social causes they care about and is cited as one of the reasons the general public is not more responsive to climate science.[48] Walter's ensuing breakdown, in which he decides he is tired of "being Mr. Good" and questions, "What if I were Mr. Bad?"[49] comes after the media realization of his moral compromises made to ensure land for the warbler park:

> . . . the Cerulean Mountain Trust was in bed with the coal industry and the defense contractor LBI, was allowing large-scale MTR [mountaintop removal mining] on its supposedly pristine reserve, was hated by local environmentalists, had displaced salt-of-the-earth country people from their ancestral homes, and had been founded and funded by a publicity-shy energy mogul, Vincent Have, who, with the connivance of the Bush administration, was destroying other parts of West Virginia by drilling gas wells.[50]

Walter's epiphany regarding his own complicity in the global dilemmas he so desperately fights to bring awareness to may serve as readers' own realization of guilt as well since we often are, as Walter, both benefactor and victim. This situation alludes to the compromises people make every day in their own efforts to contribute. It speaks to a real difficulty in defining 'the greater good' as it corresponds to personal interest in certain issues and causes.

Individual moral compromise is represented in the text by Walter through Coyle Mathis—who Walter's group finally persuades to leave his land in Forster Hollow—but the social compromise, pinned by Franzen mainly on the American middle-class, belongs to and affects all of us. As Walter exclaims,

> You, too, can help denude every last scrap of native habitat in Asia, Africa, and South America! You, too, can buy six-foot-wide plasma TV screens that consume unbelievable amounts of energy, even when they're not turned on! But that's OK, because that's why we threw you out of your homes in the first place, so we could strip-mine your ancestral hills and feed the coal-fired generators that are the number one cause of global warming and other excellent things like acid rain. It's a perfect world, isn't it?[51]

With *Freedom*, Franzen has exposed that individually and culturally Americans have avoided reaching a level of consciousness that demands accepting responsibility for the lifestyles we enjoy. As Walter devastatingly points out, it is these lifestyles—the almost vicious protection of individual freedoms—that have led to the detriment of the environment and other species. His (and Franzen's?) poignant rage is directed not only at his audience in the body-armor plant but also at readers:

> WE ARE ADDING THIRTEEN MILLION HUMAN BEINGS TO THE POPULATION EVERY MONTH! THIRTEEN MILLION MORE PEOPLE TO KILL EACH OTHER IN COMPETITION OVER FINITE RESOURCES! AND WIPE OUT EVERY OTHER LIVING THING ALONG THE WAY! IT IS A PERFECT FUCKING WORLD AS LONG AS YOU DON'T COUNT EVERY OTHER SPECIES IN IT! WE ARE A CANCER ON THE PLANET! A CANCER ON THE PLANET![52]

Walter's outburst may not be a call to action, *per se*, but it does illustrate the transformational potential of literature. As Chip Heath and Dan Heath say of statistics in *Made to Stick: Why Some Ideas Survive and Others Die*, "Statistics are rarely meaningful in and of themselves. Statistics will, and should, almost always be used to illustrate a *relationship.*"[53] As in a WW II nuclear weapons example they cite, the point of this relationship is "to hit people in the gut with the realization that *this was a problem that was out of control.*"[54] As Franzen's readers, we may not remember thirteen million, but surely we will remember Walter's identifying us as a cancer on the planet.

In *How to Read Literature Like a Professor*, Thomas C. Foster comments that "Writing that engages the realities of its world—that thinks about human problems, including those in the social and political realm, that addresses the rights of persons and the wrongs of those in power—can be not only interesting but hugely compelling."[55] Franzen's novel is most certainly compelling—as evidenced by the controversial commentary and hype in the wake of its publication. More significantly, interest in his characters may urge readers to contemplate the very real social relationships and problems they illustrate. Churchland's work provides some insight on those relationships and problems, and how they have contributed to Walter's angry realizations: "Trust can be expanded beyond the circle of kin and familiar folks if the institutional arrangements can be counted on to assure a reasonable level of trustworthiness of participants, known and unknown."[56] As Walter's goodness meets the greed of big business, readers learn how and why this trust is compromised:

> The coal companies had reason to fear that the warbler would soon be listed under the Endangered Species Act, with potentially deleterious effects on their freedom to cut down forests and blow up mountains. Vin believed that they

> could be persuaded to help the warbler, to keep the bird off the Threatened list and garner some much-needed good press, as long as they were allowed to continue extracting coal.[57]

Trustworthiness, in family and social relationships (and individual positioning in a chosen environment) comes down to how we wield personal, social, and institutional freedoms. What Franzen presents readers with, quite simply, is the domino effect: freedom impacts individuals, which affect other species, which collide with other communities, then other cultures, and globally, environments and resources.

Of course, as Joey's confused reaction to 9/11 attacks and the resulting infringement on his "birthright" of "good luck" reveals, it is not so simple when blame is hard to place, when introspection leads it back to us: "The culprit, in hindsight, seemed *almost* like bin Laden, but not quite. The culprit was something deeper, something not political, something structurally malicious, like the bump in a sidewalk that trips you and lands you on your face when you're out innocently walking."[58] Joey's maturation throughout the novel and his balance of attachment priorities (from self, to Connie and family, and to global affiliates) provides insight as to how readers might best understand and react to this "bump" in our lives, our own complicit behaviors.

Science complements a literary interpretation of this experience by illustrating the social motivations individuals experience. In the essay "Cognitive Ecocriticism: Human Wayfinding, Sociality, and Literary Interpretation," interdisciplinary scholar Nancy Easterlin reviews how our evolutionary history and neurobiology support "the development of strong emotional attachments" and our tendency to "seek a physical proximity to others."[59] Significantly, she concludes that "positive emotions associated with the social group are similar to positive feelings members have for the environment and also significantly affect group members' attitudes toward the environment with which that group is associated."[60] Thus, stake in and perceived responsibility toward our immediate environment is determined by the social attachments we form within that environment. Easterlin's point relates to Franzen's fiction, particularly *Freedom*, in an important way:

> Art is, in effect, a way of knowing and coping with the world, one that initially, perhaps, served to strengthen human groups. While its function has mutated somewhat in modern culture, art still serves as a way of coping and knowing—of exercising problem-solving skills, of imagining alternatives, of simply taking a break from the strains of daily living.[61]

This helps highlight the importance of readers' engagement with art. In the case of environmental and social justice issues presented through fiction, readers may find themselves (like Walter and Joey) contemplating how individual choices and behaviors are contributing to much larger social attitudes and consequences. If attachment to a place and its social group is strong enough, contemplation might lead to action.

In the case of Franzen's characters, readers see Walter reprioritizing his global environmental initiatives. After his disillusionment with the trust, he realizes (in our metacognitive analysis) the futility of violating Churchland's attachment paradigm. His cataclysmic meltdown must be public because he saw himself as working for the benefit of his social affiliates—an anonymous public that cannot possibly appreciate what this work has cost him. Franzen's intimation that Lalitha is killed due to her environmental activism crystalizes Walter's reprioritization. All areas of his life—self, close-kin, affiliates—would benefit from being in balance. In her engagement with a larger community and in her role as mother, Patty has also changed. By mothering children in a local school, Patty's self finds reward so that she can interact with her own children on a more adult level (thus allowing them to become their own selves). Additionally, her friendships with colleagues at the school allow her to form more genuine relationships with her and Walter's neighbors.

As Walter was catalyzed by Lalitha's death, Patty's reprioritization was catalyzed by Richard's exposure of her autobiography. Both painful, both essential. As for Richard himself, the destructiveness of his early fame on his unformed identity prevented real attachment to close-kin and affiliates. Although he adopted Walter and Patty as close-kin, it remains a parasitic relationship until the affair at Walter's rain-swept lake house. The aftermath of this affair not only causes Patty's epiphany but also allows Richard to appreciate the value of his fame through his participation in Walter's environmental activism. By exposing the autobiography, Richard acknowledges his failure to win by connecting with Patty. In doing so, he is choosing close-kin over self, valuing Walter and Patty's relationship over his short-lived hope. In the end, his new album—as a gift to Walter—symbolizes his newly found maturity and valuation of family. His work with Free Space evolved from a ploy to be near Patty into authentic participation in a prosocial cause. By the end of the novel, this family unit has reunited on a stronger, healthier, more balanced footing.

Patrick Colm Hogan's essay "On Being Moved: Cognition and Emotion in Literature and Film" addresses the importance of merging humanities with sciences through a review of how human cognition and emotion affects our reaction to literature and film. Hogan asserts that "to understand literary emotion, we need to understand emotion generally."[62] One of the reasons we "respond emotionally to literature" even though "we know perfectly well that the characters are not real" is because, according to "[a]ppraisal theorists . . . we mentally simulate the experiences of a character from his or her point of view."[63] In the case of Franzen's *Freedom*, a glaring call to action lies within this simulation: we see ourselves, our family, our neighbors, our country, our world exposed in its pages. If readers are kept "awake at night" by Walter's fear that "All the real things, the authentic things, the honest things are dying off" and relate to his comment that in our individual and social lives "we just bounce around like random billiard balls, reacting to the latest random stimuli,"[64] then like Walter we must react.

Analyzing good literature toward a level of self-reflection may help readers engage more meaningfully toward self-reflection that moves beyond initial reactions. Churchland's formulation of the attachment sequence offers a science-based approach to understanding how human morality works, and a rationale for why morality is socially advantageous. A story read with this focus prompts questions of what it means—for past, present, and future—to live morally. Like Franzen's characters who learn the responsibilities inherent in freedom, readers are invited to—in the "words of wisdom from the Class of 1920" that Patty, after a night of excessive drinking, reads on a building at Jessica's college— "USE WELL THY FREEDOM."[65] The placement of this reference recalls the moral dissolution with which Fitzgerald characterized that decade in *The Great Gatsby*. Its narrator, Nick, reflects at the end of this novel and cautions: "They were careless people, Tom and Daisy—they smashed up things and creatures and then retreated back into their money or their vast carelessness or whatever it was that kept them together, and let other people clean up the mess they had made."[66] If we are not to end up as morally despicable as they, we must be responsible where they are not. The consequence for our future, warns Fitzgerald, is "what foul dust [floats] in the wake of [our] dreams."[67] Franzen's *Freedom*, through a reading lens respecting both the sciences and humanities, shows us how to live in a way that will not corrupt our dreams.

Given this endeavor, the opening quote to *Freedom* is, appropriately, Shakespeare's:

> Go together,
> You precious winners all; your exultation
> Partake to everyone. I, an old turtle,
> Will wing me to some withered bough, and there
> My mate, that's never to be found again,
> Lament till I am lost.
>
> —*The Winter's Tale*[68]

One of Franzen's most illustrative passages suggests what we readers, like Walter and Shakespeare's turtle dove, may ultimately lament. The passage begins "It was the season of migration, of flight and song and sex" but details the increasingly arduous migration of "a few hundred bird species"[69] to the United States, making clear Americans' shared complicity for their treacherous journey and hostile destination:

> . . . Each year, they arrived to find more of their former homes paved over for parking lots or highways, or logged over for pallet wood, or developed into subdivisions, or stripped bare for oil drilling or coal mining, or fragmented for shopping centers, or plowed under for ethanol production, or miscellaneously denatured for ski runs and bike trails and golf courses.[70]

Fortunately, the conclusion of this foreboding passage offers readers hope, and, I believe, a call to responsible living: "But the United States was still a rich and relatively young country, and pockets full of bird life could still be found if you went looking."[71] This valuation is as significant as his protagonists' reconciliation—Walter eventually reunites with his mate, a reunion which directly fosters prosocial promotion within their neighborhood by allowing Walter—who, when alone, had harassed his neighbors about allowing their cats to roam freely and kill song birds—to "invite [his neighbors] to join him and a scientist friend on a series of nature walks in May and June, to know their local heritage and to see, up close, some of the marvelous life the woods were full of."[72] Also important is that when the couple designates their home and land at Nameless Lake as a bird sanctuary, they move on to another community where Walter, presumably, will continue his wildlife awareness efforts with Patty's support. As Churchland notes about the role of science in moral dilemmas, "we should be open to the possibility that a deeper understanding of the nature of our sociality may shed light on certain of our practices and institutions, and cause us to think more wisely about them."[73] Walter's environmental goals may seem small compared to his initial quest to save the world, but they are no less significant. Through this compromise, and Walter's evolving personal, familial, and social fulfillment, Franzen offers hope that readers may live responsibly by creating in our lives a more positive chain reaction: we fulfill ourselves, love our families, protect our communities, preserve our country, impact our world.

This resolution calls to attention the moral obligation as a culture to address climate change: Educational objectives must challenge individuals, families, and communities to use freedom well.

NOTES

1. Amitav Ghosh, *The Great Derangement: Climate Change and the Unthinkable* (Chicago: The University of Chicago Press, 2016), 12–29.
2. Patricia S. Churchland, *Braintrust: What Neuroscience Tells Us about Morality* (Princeton: Princeton University Press, 2011), 8.
3. Ibid., 9
4. Ibid., 8.
5. Ibid., 9.
6. Ibid., 27.
7. Ibid., 30.
8. Ibid., 31.

9. Paul Charles Griffin, "Sound and Fury, Signifying . . . What, Exactly? Notes on the Franzen Wars," *The Common Review* 9, no. 2 (2011), 15–16.
10. Jonathan Franzen, Interview with Gregg Gambina, quoted in Griffin, "Sound and Fury," 16.
11. Griffin, "Sound and Fury," 16.
12. Ibid.
13. Sam Tanenhaus, "Peace and War," *New York Times Book Review* (2010): 77.
14. Ibid., 79.
15. Philip Hensher, "Family Fallout," *Spectator,* 46 (2010): 88.
16. Leo Robson, "I Feel Your Pain: The Difficulty of Being Jonathan Franzen," *New Statesman*, 74 (2010): 90.
17. Jonathan Franzen, *Freedom* (New York: Farrar, Straus and Giroux, 2010), 21.
18. Ibid.
19. Ibid.
20. Ibid., 5.
21. Ibid., 4.
22. Ibid., 8.
23. Ibid., 66.
24. Ibid., 133.
25. Ibid., 140.
26. Ibid., 14.
27. Ibid., 21.
28. Ibid., 14.
29. Ibid.
30. Ibid., 141.
31. Ibid.
32. Ibid., 152.
33. Ibid.
34. Ibid., 153.
35. Ibid., 155.
36. Ibid., 153.
37. Ibid., 13.
38. Churchland, 31.
39. Ibid., 62.
40. Ibid., 63.
41. Franzen, *Freedom*, 158.
42. Ibid., 161.
43. Ibid., 173.
44. Churchland, *Braintrust,* 175.
45. Franzen, *Freedom*, 344.
46. V. S. Ramachandran, *The Tell-Tale Brain: A Neuroscientist's Quest for What Makes Us Human* (New York: W.W. Norton and Company, 2011), 291.
47. Franzen, *Freedom*, 481.
48. See, for example, George Marshall's *Don't Even Think About It: Why Our Brains Are Wired to Ignore Climate Change.*
49. Ibid.
50. Ibid., 472–473.
51. Ibid., 483.
52. Ibid., 484.
53. Chip Heath and Dan Heath, *Made to Stick: Why Some Ideas Survive and Others Die* (New York: Random House, 2007), 143.
54. Ibid.
55. Thomas C. Foster, *How to Read Literature Like a Professor* (New York: HarperCollins, 2003), 110.
56. Churchland, *Braintrust*, 65.
57. Franzen, *Freedom*, 210.

58. Ibid., 233.
59. Nancy Easterlin, "Cognitive Ecocriticism: Human Wayfinding, Sociality, and Literary Interpretation," in *Introduction to Cognitive Cultural Studies,* Lisa Zunshine, ed., (Baltimore: The Johns Hopkins University Press, 2010), 262.
60. Ibid.
61. Ibid., 263-64.
62. Patrick Colm Hogan, "On Being Moved: Cognition and Emotion in Literature and Film" in *Introduction to Cognitive Cultural Studies*, Lisa Zunshine, ed., (Baltimore: The Johns Hopkins University Press, 2010), 245.
63. Ibid.
64. Franzen, *Freedom*, 218.
65. Ibid., 184.
66. F. Scott Fitzgerald, *The Great Gatsby* (New York: Simon & Schuster Inc., 1925), 188.
67. Ibid., 6
68. William Shakespeare, *The Winter's Tale*, in *The Norton Shakespeare*, Stephen Greenblatt, ed. (New York, NY: W.W. Norton and Company, Inc., 1997) 5.3.131–136.
69. Franzen, *Freedom*, 485.
70. Ibid.
71. Ibid., 486.
72. Ibid., 561.
73. Churchland, *Braintrust*, 190.

Chapter Six

Ecopsychology

Harmonizing Our Paths

"'The country knows,'" a Koyukon elder warns. 'If you do wrong to it, the whole country knows. It feels what is happening to it.'"[1] In the appropriately titled essay, "Where Psyche Meets Gaia," Theodore Roszak points to this northern Alaskan tribal wisdom in his analysis of humans' profound ecopsychological relationship to Earth. His elaboration of this concept elucidates how powerfully interwoven is humanity's fate with Earth's: "ecopsychology proceeds from the assumption that at its deepest level the psyche remains sympathetically bonded to the earth that mothered us into existence."[2] This ecological perspective on the human psyche "suggests that we can read our transactions with the natural environment—the way we use or abuse the planet—as projections of unconscious needs and desires." Indicating behaviors enacted by a Taker culture, ". . . Toxic wastes, the depletion of resources, the annihilation of our fellow species," Roszak asserts that "all these speak to us, if we would hear, of our deep self."[3] As we have seen, what these actions reveal is a species that seems to have abandoned spiritual connection with the earth for a mentality of conquering it. As Wendell Berry suggests, when we begin to view Earth in this way—as a planet we can wield power over and manipulate—we essentially reverse our evolutionary connection to it as a life-giving source that provides our continued sustenance. This unhealthy psychology influences our conscious actions, or lack thereof, further distancing us even without our awareness. Restoring a more natural relationship to Earth—as both dependents and stewards—might therefore provide healing for both humanity and the planet from abuses inflicted and suffered.

As psychologist James Hillman points out in "A Psyche the Size of the Earth," this is not a new concept (at least in the history of psychology or traditions of indigenous peoples), but it is a perspective that, for the sake of human and environmental health, we can no longer afford to deny or ignore. Our healing as individuals and as a culture "requires not merely a journey to the interior but a harmonizing with the environmental world."[4] By fostering such a journey for ourselves and our young people, it is not only the environment we have a chance to save but ourselves and each other. As I have tried to demonstrate throughout, this chapter highlights a variety of resources available to inform an ecopedagogy that will reestablish our vital position as Leavers who see ourselves as part of, rather than master over, Earth's community. While, as in previous chapters, the fiction has been selected for its cultural and educational accessibility, these stories and accompanying research specifically witness humanity's deeply embedded psychological and spiritual connection to the environment as well as acknowledge our responsibility for its care.

A narrative that explicitly illuminates the significance of our ecopsychological connection to Earth and of what Nature can teach us of "our deep self" is Mary Shelley's *Frankenstein*. I reference it here again[5] because it is a classic piece included in curricula around the globe, developing an even stronger and ever-changing presence in the popular culture of each generation. This analysis considers the detrimental consequences of our disconnection from Nature as inherently related to the quality of our familial and social interactions. High school educator Gladys V. Veidemanis explains why Shelley's allusion to Coleridge's "The Rime of the Ancient Mariner" is relevant to the link: "Frankenstein, the Mariner, and Walton as well are revealed as weighted with a similar albatross of guilt as a consequence of the failure of love, a turning away from family, nature, and human fellowship that eventually condemns all three to a frigid wasteland of aching solitude and enduring regret."[6] Walton's reference to the poem in a letter to his sister does more than establish the unwelcoming setting toward which he embarks; he recognizes what he risks losing but his "love for the marvelous"[7] pushes him toward adventure nonetheless.

As with Frankenstein and the Mariner, this foreshadowing of Walton's isolation warns readers of the "frigid wasteland" symbolically awaiting those who turn away from the bonds that make us human. Ironically, it is Frankenstein's creation who instructs readers in his plea to Victor: "'Make me happy, and I shall again be virtuous'" (p.100).[8] As Veidemanis notes, "the reader is never allowed to forget that the initial wrong was Frankenstein's in banishing the monster from love and participation in the human family."[9] The price of Walton's and Frankenstein's ambition is a similar isolation to that which the creature laments, but solution—happiness through companionship—is sought only by the creature. Shelley's construction of setting further emphasizes the importance of companionship and nature on our own well-being. As Frankenstein himself confesses of confining himself to the lab:

> Winter, spring, and summer passed away during my labours; but I did not watch the blossom or the expanding leaves—sights which before always yielded me supreme delight—so deeply was I engrossed in my occupation. The leaves of that year had withered before my work drew near to a close; and now every day showed me more plainly how well I had succeeded. But my enthusiasm was checked by my anxiety, and I appeared rather like one doomed by slavery to toil in the mines, or any other unwholesome trade, than an artist occupied by his favourite employment. Every night I was oppressed by a slow fever, and I became nervous to a most painful degree; . . . the fall of a leaf startled me, and I shunned my fellow-creatures as if I had been guilty of a crime.[10]

Important to note is that Frankenstein's self-imposed isolation exists as a direct result of his attempts to outmaneuver Nature by creating an immortal. Though readers can empathize with the grief that inspires the scientist's quest—his mother's death—Shelley carefully demonstrates his deeply misguided efforts. A better vision for humanity, as Shelley imagines it, lies in our connectedness both to loved ones and to the natural world. Like Victor's creature, we innately recognize this need which is perhaps why we cannot help but empathize more with the creature, for whom it is denied, than its creator, who chooses his solitude in the name of scientific progress.

In "Frankenstein's Fallen Angel," Joyce Carol Oates reinforces why we connect with Shelley's creature—his lament echoes our own quest to understand identity and purpose, as well as our desire for acceptance and love:

> 'My person was hideous, and my stature gigantic: What did this mean? Who was I? What was I? Whence did I come? What was my destination?' . . . Surely one of the secrets of *Frankenstein*, which helps to account for its abiding appeal, is the demon's patient, unquestioning, utterly faithful, and utterly *human* love for his irresponsible creator.[11]

Interestingly, it is both his humanness and his critique of humanness which makes him a sympathetic character. Identifying and empathizing with him thus encourages readers to experience the novel with inquiry and reflection toward our own lives. Acknowledging Shelley's context, today's contemplation should occur not only in individual ambitions, but also close relationships, responsibility to society, interactions in nature, and perceived progress in science and technology.

By means of its various adaptations, Shelley's tale is one that readers of all ages and cultures can access, making its lessons widely influential. In Oates' words, *Frankenstein*

> . . . is a parable of our time, an enduring prophecy, a remarkably acute diagnosis of the lethal nature of denial: denial of responsibility for one's actions, denial of the shadow-self locked within consciousness. . . . [the creation's]

> archetypal significance rings true. 'My form,' he says eloquently, 'is a filthy type of yours.'[12]

And in fact, readers are far more disgusted by his creator than the creature himself, just as we may experience shame at the human behaviors he critiques throughout the text. Employing Shelley's remarkable symbolism to discuss our "filthy" past could compel readers to consider present parallels evidenced by environmental issues—namely the availability, use, and distribution of natural resources. How the privileged treat those who face the most damaging effects of climate change may lead to unfortunate realizations that echo the creature's.

Toward the conclusion of the novel, Frankenstein's creation draws him further into a vast wilderness as they vow to destroy one another. The creature has killed his creator's closest loved ones so that he would experience the same isolation the creature has lamented throughout the tale. The lesson Shelley conveys in both their exiles (and ultimately, their deaths) lies in the juxtaposition of their initial goals: Victor's creation seeks the companionship of family and love that Victor himself forsakes in his pursuit of science. In an important way, this scene helps highlight Shelley's construction of Frankenstein's character in light of ecopsychological concerns about our similar detachments—familial, social, environmental. Readers' reaction to this ambitious scientist is affected by a recognition of his increasing separation—from people and the natural world—and might signal cautions about our own. While the creature's happiness is denied him, Frankenstein has learned too late the consequences of trying to outmaneuver the natural laws of life and death that had taken his mother from him. The irony of his quest is that it ultimately ends all the bonds that first motivated him. Especially in today's culture that favors social media interactions over personal connections and technological advancements over basic problem-solving logic, there are many unfortunate parallels to Shelley's story. Like the explorer Walton at the close of Frankenstein's narration, readers must choose their path ahead. The hope is that we learn to choose more wisely than Shelley's scientist.

Education must play a vital role in helping to inform this choice. Advocating a transformative shift in the way we think and act as individuals and societies, education professor C. A. Bowers notes in "Changing the Dominant Cultural Perspective in Education," that "the aspects of the ecological crisis that can be attributed to human behavior are cultural in nature and directly related to the modern ideas and values that are the basis of the knowledge learned in teacher education programs and in the other areas of graduate studies in education."[13] This important statement identifies an interdisciplinary instructional source that has both contributed to, and has the power to resolve, a cultural mind-set responsible for the ecological and social dilemmas we face today. A rationale for change in the preparation of education professors but also in graduate studies generally, Bowers' explanation echoes Shelley's and Atwood's storytelling (in the *MaddAddam* trilogy) by suggesting that our present circumstances portend a future reality based on superficial, economically motivated solutions to the most quintessential human concerns:

> Today we are witnessing how science and technology are being used to transform nearly every aspect of community life—including the healing arts, play, ceremony, nurturing, mentoring, and convivial activities—into market relationships. Megastores such as Toys 'R' Us have replaced the transgenerational sharing of children's play and stories, the preparation of food and recreational activities have been turned into huge economic markets, and knowledge of self and relationships have become the domains of personal experts whose primary motivation is to sell either chemically based or trendy psychological healing strategies.[14]

If education is in many ways the root cause of a damaging cultural mind-set, then it can also be a solution. Ecocritical pedagogical shifts could support positive influences of teacher preparation on student behavior. This understanding has already begun to manifest itself in the public education system, if only on a small scale. Resulting from California's 2003 Education and the Environment Initiative (EEI) is a potential model for such partnering through a state environmental curriculum initiative. The effort has brought together various education, legal, business, and other stakeholders in impressive collaboration. As described on California's EEI website, this curriculum, currently based in science, history, and social science content standards,

> . . . is designed to stimulate critical thinking in students in an engaging manner, based on the growing body of evidence from national studies indicating that achievement increases when students learn academic content through environment-based instructional strategies. This approach makes education relevant for students, connecting their learning to the communities they live in and stimulating their involvement with the world around them.[15]

California's progress toward environmental education as well as energy efficiency goals is exciting to witness as more states consider similar change and commit to asserting their goals politically. By establishing local and state programs that address environmental issues, schools can build participation in a more global exchange of concerns and solutions. Students who learn to recognize and act on local environmental concerns will be more inclined to understand and influence them on a broader scale.

This effect begins with learners who can see the patterns that connect individual action to social and environmental consequence. In the introduction to *Ecological Literacy: Educating Our Children for a Sustainable World*, editors Zenobia Barlow and Michael K. Stone summarize their experience working with The Bay Institute of San Francisco to enhance environmental education in area schools. Attempts to address issues in local ecosystems as well as the identification of other regional problems including "malnutrition, solid waste generated by students throwing away their lunches, underachievement, and vandalism"[16] led to significant conclusions for education practices. Echoing Bowers' concern, they report that, "we recognized them not as isolated problems, but parts of one overarching problem of *disconnection*: of rural communities from urban life, of food from people's understanding of its origins, of health from the environment—and of problems from the patterns that perpetuate them."[17] Acknowledging this disconnection and targeting these patterns can start the cultural healing process necessary to address many of our common problems. For example, empowering our schools as a means of reconnecting urban lifestyles with rural philosophies regarding the earth can foster individuals' appreciation for where and how food is grown. Regardless of geographic location or economic status, making these connections could motivate social groups to demand better. Imagine how much more powerfully productive our workforce would be if every member had learned at an early age to respect the relationship between individual and environmental health—if every factory worker, scientist, farmer, janitor, chef, doctor, architect, transportation worker, chemical manufacturer, and so on, knew the explicit connection between their health and the health of the environment. What ethical and moral implications could there be? Consider the impact these conscientious members could have on decision making in the larger organizations and corporations for which they work, or on the governmental policies for which they vote.

Popular children's literature can help even young children understand these connections. With the March 2012 film rendition of Dr. Seuss' classic cautionary tale *The Lorax*, young audiences were reminded of its lesson for humanity, but even much older ones could still relate its simple message from reading the book as children: When we abuse our natural resources, and ignore the warnings we are offered, they may be lost forever. The story's beginning lines invites readers or audiences to connect present with past as the Once-ler will do for the boy who seeks him out: "'Now I'll tell you,' he says, with his teeth sounding gray, 'how the Lorax got lifted and taken away. . . It all started way back . . . such a long, long time back. . . .'"[18] In the flashback of events that unfold, Seuss' Lorax tries to show the greedy Once-ler that we should not ignore the harmful impact of our behavior toward the earth and each other. The message is instructional in its appeal to restrain abuses of natural resources for material gain by gradually revealing the impact of the Once-ler's choices—on the Truffula Trees, the Bar-ba-loots that depend on them, the air that is polluted by his Super-Axe-Hacker, the waste produced by his Thneed factory, and the various creatures forced to leave the destroyed environment. Only when his family and the cautionary Lorax himself has gone does the Once-ler begin to understand what he had done. On his path to success and profits, he may not have meant to destroy the special trees that he used for raw materials but destroy them he did. The wasteland that remains reflects the many barren landscapes left in the wake of our own attempts at "biggering." In a way that they can relate, *The Lorax* asks children to demonstrate compassion for all inhabitants of Earth because of the unique beauty contributed by each. The sweet story moves beyond sentimentality in the face of unprecedented documentation revealing how humans have disrupted the harmony of the world by destroying what makes it beautiful. Like the Once-ler who learned too late what he had done, we are only beginning to understand the resulting damage to Earth and to our psyches.

I have listened to older students easily and excitedly recount the environmental messages of this favorite Dr. Seuss tale. Imagine how transformative their reading experience could have been if their love for the story had been nurtured with hands-on lessons about the use of resources in their own communities. How concrete would be the warnings of the Once-ler's experience if young readers were shown the true costs of the products they loved, if we could appeal to their consciences before they became active consumers themselves? Consider the results if every young reader in America learned responsibility by being taught how to replenish what was taken—how to, as the Once-ler eventually directs the boy in the story to do, "grow a forest" and "protect it from axes that hack."[19] To make simply this one message an impetus for national action, we need only invite every kindergartener in America to leave the classroom for a day and plant a tree for the Once-ler, who finally learns what the Lorax meant by his closing caution: "unless."

At all levels of education, students' ability to identify, understand, and influence the resulting pattern of consequences that stem from disconnection with nature will likely be determined by how effectively educators can connect learning across subject areas. Apart from educators who understand the value of (and are willing to invest time in) interdisciplinary collaboration, our system is largely structured by subject division—in elementary, middle, and high schools, the bell schedule, rather than any thoughtful organization of learning, determines students' focus area, often down to the minute. In higher education, students ultimately design their experience based on interest or ability, choosing what knowledge to include or exclude in their career pursuits. With an imposed structure that artificially separates learning experiences or allows students to do so themselves, how do we hope for them to recognize, much less address, the interrelated outcomes Barlow and Stone discuss? As the premise of my argument asserts, in the consequence of our most complex human behavior patterns—climate change—we need young people to not only see but comprehend and react to these corollaries. We cannot so consistently compartmentalize education and then expect students to magically connect the dots when they leave the classroom. The World-

Watch Institute's publication, *State of the World 2013: Is Sustainability Still Possible?* explores a compelling rationale as to why our survival as a species depends on their ability to do so. In a chapter titled "Moving Toward a Global Moral Consensus on Environmental Action," philosophy professors Kathleen Dean Moore and Michael P. Nelson (editors of *Moral Ground*) expound on why a specific collaboration between sciences and humanities is our best hope of mitigating effects of climate change:

> Indeed, scientists have heroically expanded knowledge and explained it to the public on the assumption that if people only knew, if they only knew, then they would act. This, unfortunately, is a fallacy. Better to say, if people only knew the facts about the harmful effects of climate change on the human prospect, and if they affirmed basic principles of justice and compassion, then they would act. It is from the partnership between science and ethics that policies are born. For this reason, university departmentalization and the myriad isolations of expertise, science/religion divides, and other forces that weaken the connection between the realm of the first premise (generally science and technology) and the realm of the second premise (literature, art, religion, indigenous wisdom, ethics, history) have made it harder to devise effective policies.[20]

Given the ecological dilemmas humanity currently shares, our educational approach must be shaped to engender critical thinking that examines these areas as crucially connected rather than distinctly separate. In the case of climate change, we need stories that help us face the science.

Professors Richard Beach, Jeff Share, and Allen Webb present excellent examples of using climate-related stories from a variety of genres and in diverse formats in their recently published instructional resource, *Teaching Climate Change to Adolescents: Reading, Writing, and Making a Difference*. They urge educators to commit to incorporating climate change in their instruction because "our students inherit and will live their lives in a climate changed world. They deserve to understand their world deeply and meaningfully, to consider large social and cultural values, ethical questions, and ways we can transform how we now live in order to save life on Earth."[21] As they note of emerging literature that supports this effort, "climate fiction foregrounds the impacts of climate change on individuals and societies in ways that are meaningful to readers. It supplements factual/scientific information to create richer and more balanced understanding of the future. . . . readers can see multiple natural and human systems interacting and consider the consequences."[22] As with all literature, readers should be encouraged to explore not only interactions within the stories themselves but also within the very real circumstances that may have inspired them—situating the authors as witness to their changing climate and the stories as cultural artifacts. More advanced students may be asked to consider the recent proliferation of such stories in relation to Amitav Ghosh's criticism in *The Great Derangement* that so few authors have stepped forward in significant ways. Examining artists' silence on the issue could lead to interesting conversations about our own.

Conservation Biologist David Ehrenfeld critiques higher education's lack of response to climate change in *Becoming Good Ancestors: How We Balance Nature, Community, and Technology.* As a member of the environmental education grant organization Educational Foundation of America, Ehrenfeld laments "the extreme scarcity of exciting, innovative, useful proposals coming out of the major research universities: Harvard, Princeton, Stanford, Berkeley, and the like," noting that "the problems that the universities are doing little or nothing to address—either in teaching or in research—are those that we must confront if our civilization is to prosper. . . ."[23] He also suggests there is a lack of service learning at universities, which further limits their capacity to address environmental issues. Where such programs do exist, they are often not connected to academic programs, reflecting a lack of comprehension as to the scope of current and impending problems. Ehrenfeld echoes many others in urging education reform at all levels, including helping to "make kindergarten to twelfth-grade teaching an attractive profession" and providing funding for environmental studies, field trips, and projects:[24]

> If we don't do something on a grand scale and do it quickly, the future adults who are now four, five, six, and seven years old are going to be incapable of saving their country when they grow up, and incapable of benefiting from a rigorous college education. They need while they are young the real experience of nature, and the real experience of group (community) learning and cooperation, and the real experience of the joy of sustained and satisfying work.[25]

The experiences he suggests can nurture meaningful ecopsychological connections by encouraging students' connection to the outdoors as well as with one another through the projects in which they engage. In *Occupy Education: Living and Learning Sustainability,* Tina Lynn Evans explains why service learning specifically helps promote prosocial behaviors that extend beyond the school experience and into communities:

> For many students who have grown up in suburbia not even knowing their neighbors, involvement in community work of any kind may be unfamiliar and somewhat personally challenging (see Loeb, 1999). Service learning, a pedagogical method that includes conceptual classroom learning, action in the community that is related to conceptual learning, and structured reflection on the relationships between classroom and experiential learning, appears to be effective in helping students overcome this hurdle.[26]

With appropriate guidance and community support, service learning objectives and projects can be implemented at all levels of education and be as simple as trash cleanup or as involved as an internship with a Community Supported Agriculture (CSA) group. Opportunities to volunteer with and learn from various community members benefit all involved by potentially creating long-term cycles of supportive participation. When aligned effectively to academic experiences, service learning also correctly emphasizes the teacher's role as facilitator, awarding more responsibility for learning and continuous participation to the students themselves.

Along with considering how and what we teach, getting students outside is an important change we can make, especially considering the screen time students engage in daily on top of the sedentary, indoor experience of most school settings. Paul Shepard, one of the first to work toward an ecopsychology that respects "the interplay between human psychology and humankind's increasingly destructive environmental behavior,"[27] investigates the role of nature in a developing human psyche. In "Nature and Madness," he comments that, "even as socially intense as we are, much of the unconscious life of the individual is rooted in interaction with otherness that goes beyond our own kind, interacting with it very early in personal growth, not as an alternative to human socialization, but as an adjunct to it."[28] Like other environmentalists, philosophers, and now psychologists, Shepard examines the historical breakdown of our connection to this "otherness," adding to the discourse on current culture's disconnect (from earth, environment, plants, animals, and so on down to the microbial level). Shepard links consequential deterioration of the human psyche to modern society's mind-set and behaviors by explaining how our evolutionary history instinctively connected us to the natural world:

> An ecologically harmonious sense of self and world is not the outcome of rational choices. It is the inherent possession of everyone; it is latent in the organism, in the interaction of the genome and early experience. The phases of such early experiences, or epigenesist, are the legacy of an evolutionary past in which human and nonhuman achieved a healthy rapport. Recent societies have contorted that sequence, have elicited and perpetuated immature and inappropriate responses. The societies are themselves the product of such amputations, and so are their uses and abuses of the Earth.[29]

Put simply, Leavers enjoyed this "healthy rapport" with nature but Takers have systematically destroyed it. The goal in developing public education curricula is thus not simply to teach outdoors but to design an educational experience that arouses an inherent sense of connection with Earth and all its "Otherness." We want to encourage students' active reclamation of a physical, psychological, and spiritual bond with Nature.

A resource for educators interested in framing such an approach to reform is *Teaching Environmental Literacy: Across Campus and Across the Curric-*

ulum. Professor of Biology Heather L. Reynolds describes the collaborative effort of the campus community to develop sustainability education at Indiana University Bloomington. In her overview for the book, Reynolds explains how contributing members began to frame environmental literacy goals around that which connects us to the environment:

> We identified three central organizing themes for environmental literacy: ecosystem services, ecological footprint, and sustainability. We propose that, collectively, these three themes represent three essential elements of the human-environment interaction that every citizen should know: that humans ultimately *depend* on the environment for essential ecosystem services that support the human economy and social well-being; that the scale and scope of human activities makes humanity the dominant biological force on earth, and this domination must be appreciated at least in part for the risks it poses to the ecosystem services on which humans depend; and that a powerful alternate paradigm to domination is sustainability, through which human *alliance* with the environment can permit enduring and resilient human and natural economies.[30]

Throughout the text professors from various disciplines offer examples of how they have included these organizing themes in their courses and as part of their overall pedagogy. A common thread among the professors' contributions emerged: fostering a sense of place as a means for making the three organizing themes immediately relevant to students' lives. As they note, this is particularly important at the university level where students are likely to be from diverse backgrounds and geographic locations. As Scott Russell Sanders, professor of English, explains:

> . . . one crucial need is to recover an intimate knowledge of, and affection for, our home ground. Every neighborhood, every watershed, every wild land needs people who feel responsible for that place, who know its human and natural history, who speak resolutely on behalf of it. Whatever else our students may learn from us, they should learn the pleasure and value of such belonging.[31]

At university level, this engagement with place might be a natural extension for students who already acknowledge an attachment to their home towns or cities; for some, campus life may be the first time they will experience a true sense of place and belonging.

An early education that seeks to motivate such empathy by incorporating place-based instruction and experiences might nurture a better sense of awareness later in life. The popular 2012 film rendition of Tolkien's *The Hobbit*, a novel commonly taught in middle or early high school, offers an intriguing parallel. Students who study Bilbo Baggins, the hobbit who initially declines a chance at adventure out of love for his cozy home's security in the Shire, might appreciate the lesson he eventually learns. We understand his growth when Bilbo confidently responds to Thorin Oakenshield questioning his place on their journey: "I know you doubt me. I know you always have. I often think of Bag End. That's where I belong. That's home. You don't have one. It was taken from you but I will help you take it back if I can."[32] The dwarves' plight—being driven out of their ancestral lands by a greedy dragon and other vicious enemies—provides a motivating metaphor for empathizing with real life counterparts who face the threat of rising sea levels or extreme weather making their homes uninhabitable. It might enhance students' awareness of climate refugees, witnessed in more drastic numbers each year, and help them appreciate the layered political, social, and economic implications of environmental situations that force migration. Acknowledging with our students an appreciation of homes—our own and those of others—can lead to transformative experiences both in and out of the classroom. For Bilbo, the love of comfort and security ultimately moves him to help secure the same for others. For our students, we might employ similar motivations by engaging in active community research (including those more overtly effected by climate change) regarding the history of where we build homes, the development of towns or cities, the conditions of natural landscapes, the services offered (or not) to assist struggling community members. A class could be divided into research teams with the goal of discovering and synthesizing information about a particular aspect of a shared community. With strong facilitators—teachers, parents, community members—students might be encouraged to analyze their findings and determine what their community does really well or where it might improve. An extension of this learning process could include an actual or virtual visit to another community in effort to share successes and learn new strategies. The learning process is thus informing future decisions that students will make as denizens of communities and citizens of countries, many of which will be increasingly challenged by climate crisis and an influx of climate refugees.

The challenge—at any level of a student's education—will be to cultivate one's sense of place into an appreciation of others' sense of belonging to another place. Educators Dilafruz R. Williams and Sarah Taylor describe a middle school model for educational reform that fosters good inhabitance in their essay, "From Margin to Center: Initiation and Development of an Environmental School from the Ground Up." A California school they helped develop—Environmental Middle School (EMS)—grounds their curriculum for grades 6, 7, and 8 in key environmental concepts which help focus students on actual problems in their local environment. The structure of the school works to support meaningful connections in all subject areas; in addition to outdoor environmental and community service projects, "a variety of people from the community come to the school to present topics related to environmental, health, and other issues."[33] The school also invites elderly members of the community to come to student assemblies and share the "history of place through their stories."[34] What is exciting about this school model is that administrators and teachers are collaborating with each other and with members of their school community in an effort to raise more environmentally conscious children. The authors note that "more than fifty percent of parents and a large number of elders from the community are actively involved,"[35] reinforcing the positive community response to this approach.

Environmental Middle School centers their curricular efforts and social structures on relationships to food and community: "once a month, a class takes on the responsibility to make a community meal for the whole school. Students, teachers, and parents plan these special meals, often shopping at a nearby Nature's store that sells organic foods."[36] Students also participate in a "Salad Day" project which connects them with area farmers who teach them the basics of growing food from seeds. As we saw in chapter five, in addition to hands-on experience with sustainable agricultural practices, this kind of project introduces students to the importance of protecting and maintaining local farming, even (and perhaps especially) in urban areas. By extension, their intent is that the "EMS community will appreciate organic farming and harvesting and using food grown locally instead of that shipped thousands of miles before it gets to our tables."[37] This middle school achieves a very basic awareness for students about organic and local food through community supported agriculture; their hands-on participation in school will hopefully translate into responsible decision making as they become consumers. Thanks to grant opportunities, forward thinking communities and school leaders, there are now several models for introducing food-based education programs that include connections with local farmers and restaurants. In Alice Waters' words, "How we eat can change the world."[38]

For schools interested in developing locally relevant curricula, a concept-based interdisciplinary example is offered by Environmental Middle School, which integrated a thematic semester based on the theme of rivers:

> Besides studying their local watersheds, students picked a river in the United States and conducted historical research on the impact of that river on humans and also ways in which humans had changed the course and nature of the river. They read novels in which rivers play a significant role, did art projects and painted murals, sang songs about rivers, conducted experiments and studied water properties. . . Students wrote poems, which were compiled in a book, *A River of Words*.[39]

To build on this curricular concept at upper secondary levels in English and history, teachers might easily focus instruction through more complex river themes in some of the most commonly taught selections. A few examples include Hemingway's *In Our Time*, Conrad's *Heart of Darkness*, Morrison's

Beloved, Twain's *Huckleberry Finn*, Kingsolver's *The Poisonwood Bible*, Hurston's *Their Eyes Were Watching God*, Theroux's *The Mosquito Coast*. By thematically connecting literature through the environmental landscape and cultural symbolism of rivers, teachers encourage far more opportunities for exploring related historical, social, spiritual, and political concepts as well.

Author and journalist Richard Louv's *Last Child in the Woods: Saving Our Children from Nature-Deficit Disorder* offers compelling evidence for why we should support this kind of education. In his review of America's transitions from the first to second and third frontiers (the third marking the experience of today's children), Louv notes the significance of diminishing family farms as part of our cultural disorientation: "baby boomers—those born between 1946 and 1964—may constitute the last generation of Americans to share an intimate, familial attachment to the land and water."[40] For a culture that has largely been defined by the promise, gift, and plenty of its land, this third frontier presents a threat to our very identity—from the beliefs and values that motivate us to the consequences that await us. As Louv summarizes, trends indicative of today's generation of youth are further testament to our disconnection as an underlying source of current ecological problems:

> . . . a severance of public and private mind from our food's origins; a disappearing line between machines, humans, and other animals; an increasingly intellectual understanding of our relationship with other animals; the invasion of our cities by wild animals (even as urban/suburban designers replace wildness with synthetic nature); and the rise of a new kind of suburban form.[41]

Referring to this as the "de-naturing of childhood," Louv cites research on the "causes, extent, and impact" of this disconnection from the natural world. Of interest is a study conducted by Sandra Hofferth. She reports that "from 1997 to 2003, there was a decline of 50 percent in the proportion of children nine to twelve who spent time in such outside activities as hiking, walking, fishing, beach play, and gardening."[42] This alarming statistic is linked to a growing body of evidence that directly correlates our isolation from nature to physical, mental, and emotional problems among children and adults. Fortunately, as Louv points out, this is still a curable dilemma over which awareness and education might have some sway. The presence of nature has repeatedly shown health benefits for people of all ages, so our engagement with it at any stage of our lives can be restorative.

Of specific interest for educators and parents are emerging data in the field of ecopsychology which highlight Nature's potential role in treating common medical problems for school age children. According to Louv, "studies suggest that exposure to nature may reduce the symptoms of Attention Deficit Hyperactivity Disorder (ADHD), and that it can improve all children's cognitive abilities and resistance to negative stresses and depression."[43] Additionally, Louv reports that "a widening circle of researchers believes that the loss of natural habitat, or the disconnection from nature even when it is available, has enormous implications for human health and child development. They say the quality of exposure to nature affects our health at an almost cellular level."[44] An important facet of this relationship is the apparent link to human behavior, particularly in urban settings: "long-standing studies show a relationship between the absence, or inaccessibility, of parks and open space with high crime rates, depression, and other urban maladies."[45] In contrast, horticulturalist Charles Lewis' work reveals the cooperative benefits gained through "the palliative and socializing effect of tending plants."[46] Meg Daley Olmert elaborates on these benefits in *Made for Each Other*: *The Biology of the Human-Animal Bond:*

> In the 1960s [Lewis] used [gardening] to ignite a sense of community in the midst of poverty and rubble. He helped clear patches of NYC's Spanish Harlem to create neighborhood gardens. In precincts rife with crime, the police were amazed when these gardens were respected and allowed to grow. They were even more amazed to learn the vandals had become guardians of these tiny Edens.[47]

Interestingly, his work also demonstrated the aesthetic concern adopted by people caring for these neighborhood gardens. The appearance of buildings in the project's vicinity improved as people painted and cleaned surrounding areas. Individual contributions combined to improve a shared space that people from all backgrounds and interests could tend and value.[48] Like Lewis, we might surmise "that gardens make good neighbors. Growing beautiful and delicious things where nothing existed before brings people together and gives them a sense of control in their lives."[49] An accessible text for exploring this concept through fiction with younger readers is Paul Fleischman's *Seedfolks*. The story is set in Cleveland, Ohio, and follows the stories of multiple characters—all immigrants who live in the area but are not connected to one another—as they begin to interact amongst a space in which a young girl has planted lima bean seeds. As the garden expands and transforms the ugly space it inhabits, so too do readers witness the emergence of community and trust among these strangers. It is a moving example of how reclaiming ecological knowledge can help heal urban spaces as well as ourselves. A few similar examples of children's literature exploring this theme include Dyanne DiSalvo-Ryan*'s City Green*, Peter Brown's *The Curious Garden* and Carol Lexa Schaefer's *The Children's Garden: Growing Food in the City*.

What potential exists if school and community gardens were established on a wide scale across this country? The effort could develop bonds between young people who feel a sense of accomplishment for improving the way their community interacts with each other. The momentum for this already exists as schools all over the country establish campus gardens and cities transform open spaces into gardening plots or pollinator zones for bees and butterflies. Some progressive cities, in the absence of available land, have been creative with rooftop or vertical gardens, even instituting community composting services to support them. This not only encourages repurposing of organic waste (which otherwise just rots in landfills) but provides a simple way to contribute to a worthy environmental and social project. Fostering value for this type of participation through a shared ecopedagogical lens could impact our cultural well-being on an impressive scale.

Homer's epic poem *The Odyssey,* one of the most commonly taught titles in public high school, helps illustrate the healing power of restoring our connection to the earth. In a chapter titled "The Body and the Earth," from *The Unsettling of America*, Wendell Berry suggests that *The Odyssey* "is in a sense an anti-*Iliad*, posing against the warrior values of the other epic—the glories of battle and foreign adventuring—an affirmation of the values of domesticity and farming."[50] Consider this line, for example, which connects son to father and father to land: "Odysseus found his father in solitude / spading the earth around a young fruit tree. / He wore a tunic, patched and soiled, and leggings—/ oxhide patches, bound below his knees / against the brambles . . ."[51] The peaceful simplicity of this scene in contrast to the violent angst that characterizes much of both poems is poignant. As Berry interprets of Laërtês the king surviving as a peasant, ". . . the appropriateness of what he is doing is never questioned. In a time of disorder he has returned to the care of the earth, the foundation of life and hope. Odysseus finds him in an act emblematic of the best and most responsible kind of agriculture: an old man caring for a young tree."[52] Signifying an evolutionary, ecopsychological connection to the earth, the scene itself is symbolically instructional for the son who seeks him. Particularly for protagonists who have spent so much time away in distant wars, this moment encourages reflection on the restorative nature of landscapes one considers home. Educators might pair this concept of finding order in the simplicity of the earth with any number of current issues that suggest our disavowal of it. Reminding us how intimately our identities are tied to agriculture, director of the White Earth Land Recovery Project Winona LaDuke asserts this connection in a plea directed at preventing genetic engineering and patenting of North American wild rice:

> . . . the wild rice harvest links a traditional Anishinaabeg or Ojibwe people to a thousand years of culture and the ecosystem of a lake in a new millennium. This cultural relationship to food—manoomin, or wild rice—represents an essential part of what we need to do to repair the food system: We need to recover relationship.[53]

Her sentiment echoes that of Berry's machine metaphor, critiquing a mindset that believes we can manipulate Nature by "improving" it. Monoculture super seeds are not the answer when we need children who know how to grow food. Educating instead for the relationship recovery she calls for places cultural well-being above artificial panaceas, health outcomes over economic pursuits. It says we value students beyond their potential as consumers. This inherent connection (or disconnection) to our identity extends to other sources of earth-providing sustenance as well—particularly animals. As we saw with Atwood's character Jimmy/Snowman, there are consequences for treating animals like machines, and a society that is increasingly distanced from those raised for meat is in danger of doing so. In contrast to the way most consumers obtain their meat, Canadian Inuit activist Sheila Watt-Cloutier emphasizes the value of hunting culture in her essay "The Inuit Right to Culture Based on Ice and Snow." Her assessment highlights how understanding the value of an animal's life and death can directly benefit one's personal character:

> . . . it is not just about the killing of animals. It is about teaching our young children, our children, about the opportunities and challenges of life on the land, but also about teaching them very transferable skills, such as patience, courage, how to be bold under pressure, how to withstand stress, how not to be impulsive, how to have sound judgment and, ultimately, wisdom.[54]

As Watt-Cloutier suggests, these are transferable skills that directly benefit communities of people who must manage to live peacefully among each other. Hunters who respect the animals and land they hunt by not taking either for granted also play a significant role in conservation and other preservation efforts across the country. In the conclusion to her essay, Watt-Cloutier laments the impact of dominant culture being disconnected from nature: "Is it not because people have lost that connection between themselves and their neighbors, between their actions and the environment, that we are debating this issue of climate change in the first place?"[55] Her point is well-taken as we consider how education can help navigate a more prosocial approach to the problems that plague us.

Cultural historian Thomas Berry, in *The Great Work: Our Way into the Future*, writes that "indigenous wisdom is distinguished by its intimacy with and participation in the functioning of the natural world."[56] He describes at length how this intimacy creates spiritual understanding in these cultures. Wisdom is maintained and passed on through generations with an inherent practicality not present in the education of modern civilizations. Through experience more than instruction, children of indigenous cultures learn from an early age how to interact with the natural world and with one another. They are more inclined, therefore, to appreciate their place relationship within the biosphere. Paul Hawken similarly suggests that what indigenous cultures possess that most industrialized populations do not is a sense of the "local, intimate, familiar"[57] amidst a world that is distant and impersonal. The implication is that when we are not rooted in a place or committed to its people, we are likely to be less conscious of how our choices impact justice both for society and the environment. Because we do not consciously experience the impact of our choices, we can make them more cavalierly. In America, we are conditioned to view freedom as a positive, which in this case it isn't because we feel deceptively relieved of the responsibility for consequences. A better sense of how our choices impact the land and people immediately around us might lead, as in the case of indigenous groups, to respectful practices regarding that land and those people. Founder of the cultural institution Heyday, Malcolm Margolin relates a concrete example of the difference between modern and indigenous education in "Indian Pedagogy: A Look at Traditional California Indian Teaching Techniques." He presents the benefits of Indian Pedagogy as compared to common instructional practices in the United States through a friend's response regarding a roundhouse built in the Chaw'se State Park by the Sierra Mewuk community:

> 'We could have built it so that it would last a hundred years. But we didn't. We didn't because there was another law we had to follow. When you build a roundhouse you construct it so that it falls apart every twenty years. That way every generation has the chance to rebuild it. Every generation has to learn the songs, the ceremonies, the techniques. If you want to make a building last, you do it one way. But if you want to make the knowledge last, you do it another way.'[58]

As this example illustrates, our current pedagogy (content aside) is inept because it is largely based on instruction rather than experience. Similarly, our cultural paradigm is flawed because it seeks artificial fixes for individual and social problems that require much more thoughtful remedies. In other words, we should not be so focused on what the answers are that we forget to ensure students know how to determine them. As Margolin learns from his friend, we need to give young people a chance to rebuild.

Our collective effort to give them this chance in the face of climate change is essential. As Bill McKibben laments in his 2005 introduction to *The End of Nature* (1988), "it is the contrast between the pace at which the physical world is changing and the pace at which the human society is reacting that constitutes the key environmental fact of our time."[59] A compelling rationale for developing a citizenry that is not concerned with geographic or legal boundaries but with preservation of a shared planet, his work sharply documents our dangerous inaction:

> When I say 'nature,' I mean a certain set of human ideas about the world and our place in it. But the death of those ideas begins with concrete changes in the reality around us—changes that scientists can measure and enumerate. More and more frequently, these changes will clash with our perceptions, until, finally, our sense of nature as eternal and separate is washed away, and we will see all too clearly what we have done.[60]

How else but through widespread education reform do we efficiently alter this trajectory already in motion?

Discussing potential future commodities of technology in *Enough: Staying Human in an Engineered Age*, McKibben makes the argument for why our identity—as individuals and as a culture—relies on each of us and society to make this change in mind-set quickly. As Orr has suggested of misguided higher education concerns, McKibben fears that the path of our science is developing too far in advance of our logic. What we are capable of technologically, he explains, is daunting when we consider what is appropriate for us socially. In his discussion of our retaining our own humanity in an age of genetic engineering—where we may soon be able to choose the traits of our offspring as we essentially choose those of our pets—McKibben poses important questions about how individuality does, should, and will affect our humanness. Discussion in the chapter titled "Too Much" focuses on the differences between somatic gene therapy and germline genetic engineering; the science is complex and, as others have also cautioned, the consequences are significant. Briefly, what may seem appealing—the use of somatic gene therapy to manage or cure diseases, for example—may not be acceptable when we apply the science to manipulate matters of choice—whether our offspring are particularly athletic or intelligent or beautiful, for instance. In a society where some parents can provide the newest innovative gene selections for their offspring, McKibben questions how those offspring will understand individuality if their personalities and abilities have been preselected. Even more provocative, he questions what might happen to those offspring whose parents could not provide such an edge. Might they become our throwaway children? How will they compete in such a society?[61] Echoing themes in Atwood's and Shelley's fiction, what we are left with is the daunting proposition that a society of genetically engineered humans—a posthuman world—is currently held back by ethics rather than science. Because we have not yet allowed it to happen is not to say that it cannot or will not.

McKibben's analysis highlights Michel Houellebecq's *The Elementary Particles* as popular French fiction that "chronicles the rise of a posthuman species that has transcended mankind's tragic flaws" and corresponds to the "idea that we should be radically reconfigured or, better yet, surpassed by some wiser race."[62] A more dystopian vision is offered in Atwood's *Madd-*

Addam, in which readers witness gangs of humans viciously competing for resources and being terrorized by genetically altered, dangerous animals. Atwood's final installment to the trilogy follows a small group of vegetarians who seek spiritual reconnection with the earth, which they now share with the Crakers—peaceable posthumans created out of a motivation similar to Houellebecq's. Both authors, as well as many others, envision the fall of humanity as the self-destructive, inevitable fate of a selfish species that exercises little or no regard for other species or the earth.

We have heard the meme: "That which cannot endure, won't." It is easy, in fact, when we immerse ourselves in statistics and sciences of the myriad environmental crises for which humans are directly responsible to perceive our end as the only plausible outcome. The transhumanist movement declaims precisely this but, as McKibben is careful to note, environmentalists are at least partially responsible as well: "the movement to value everything else on earth has often talked carelessly about people, spreading the idea that we are a grim and uncontrollable race, a cancer cell metastasizing unchecked across the defenseless fabric of nature."[63] Nodding our heads about this on both levels is likely—the better nature of humans is often excluded when we learn about the damage we have caused, and we can indeed be a fierce enemy to Nature, which has been irreversibly altered by us. We have done so on such a scale that it is, in fact, difficult to separate what is natural from what has been altered by humans. The consequences of this tinkering have yet to be fully realized and are part of the debate for scientists, environmentalists, politicians, and educators alike.

However, as with characters in good literature, there is hope for redemption in us too. Like in Atwood's *MaddAddam* trilogy, there are those of us who despite all the damage we have done—to the environment, to each other, and to other species—believe we also have the capacity to do better. As McKibben challenges,

> We could drop the misanthropy and look at ourselves with clearer eyes. Yes, we've damaged the environment, we've enslaved our fellow man, we've slaughtered on a vast scale. This is the truth. But it's not the whole truth, or even the main one. To put it bluntly, *the fight to ward off a posthuman future begins with at least a muted celebration of the human present.*[64]

As an educator of young people, this allowance feels right. For all the darkness of human nature in life and in literature there is also kindness, compassion, love, and responsibility. We can change for the better. Like the best villains of fiction, humans are not entirely malevolent and, as McKibben argues, there is a case to be made for saving us, too:

> Macbeth is our archetype of the man whose ambition leads him to reject the natural order, in his case by killing the king. The consequence is that he loses

> all that is meaningful in life—the blessing of sleep, trust in others, the comfort of family, the power of compassion.[65]

These need not be our consequences. When students study this play they are often so immersed in the evil of Macbeth's (also his wife's) persistent scheming and unchecked greed that they need to be reminded of the audience's first encounter with him, where he proves a valorous, loyal soldier to his king and country. For Macbeth, students learn, conscience evolves too late. Perhaps ours will catch on. Many authors and scientists present a similar caution with regard to human behavior toward the environment and each other. If there is any hope for a redemptive outcome to the climate crisis assuredly under way, we must meet the facts with reflection and action rather than denial and dismissal.

Regarding the ethics and consequences of a looming posthumanism as solution to these crises, McKibben appeals to our ecological conscience in a chapter appropriately titled "Enough":

> Ecology, the great emergent science of the twentieth century, has tried to understand the incredible complexity of relationships that govern the living earth—tried to understand, among other things, how we might fit. But if we finally and forever breach all those relationships, emerge as a limitless force, then what possible interest will any of the rest of life hold?[66]

In other words, if we envision our future as a genetically altered species superior to all others, what meaning will we find in this place? Forsaking our natural place in the order of things, as Macbeth learned, may leave us wishing too late we had chosen a different path. And so our "*celebration of the human present*" should perhaps not be so "*muted.*" If we care more about how we "fit" on this earth than about how we might outsmart it by genetic manipulation, we ought to heed a multitude of warnings. We should not pursue the chimera of technological solution; we must accept that the work of rebuilding will be arduous but may also prove redemptive.

We change the well-documented dismal ecological trajectory we currently face as a nation and world by educating a citizenry that is ecologically conscious and socially proactive. We inform a morally conscientious consumer culture that drives American economy in its demand for environmental and social justice. We heal ourselves and the planet by reestablishing ecopsychological connections that honor our relationship to the natural world and all its inhabitants. Should we doubt this path, we must return to the words of biologist and writer Rachel Carson, who perceptively weaves science and literature in her frightening opening to *Silent Spring*, "A Fable for Tomorrow": "No witchcraft, no enemy action had silenced the rebirth of new life in this stricken world. The people had done it themselves."[67]

Let's keep reading.

NOTES

1. Theodore Roszak, "Where Psyche Meets Gaia," in *Ecopsychology: Restoring the Earth Healing the Mind*, ed. Theodore Roszak, Mary E. Gomes, Allen D. Kanner (San Francisco: Sierra Club Books, 1995), 6.
2. Ibid., 5.
3. Ibid., 5.
4. James Hillman, "A Psyche the Size of the Earth," in *Ecopsychology: Restoring the Earth Healing the Mind*, xix.
5. See also chapters four and five.
6. Gladys V. Veidemanis, "'Frankenstein' in the Classroom," *The English Journal* 75, no. 7 (Nov. 1986): 63, http://www.jstor.org/stable/818504.
7. Shelley, Mary. *Frankenstein with Related Readings* (New York: Glencoe McGraw-Hill, 2000), Letter Two.
8. Ibid.
9. Ibid.
10. Shelley, Mary. *Frankenstein with Related Readings* (New York: Glencoe McGraw-Hill, 2000), 56.
11. Joyce Carol Oates, "Frankenstein's Fallen Angel," *Critical Inquiry* 10, no. 3 (Mar. 1984), 545, http://www.jstor.org/stable/1343306.
12. Ibid., 553.
13. C. A. Bowers, "Changing the Dominant Cultural Perspective in Education," in *Ecological Education in Action: On Weaving Education, Culture, and the Environment*, ed. Gregory A. Smith and Dilafrus R. Williams (Albany: State University of New York Press, 1999), 161.
14. Ibid., 164.
15. California Education and the Environment Initiative, www.californiaeei.org.
16. Michael K. Stone and Zenobia Barlow, *Ecological Literacy: Educating Our Children for a Sustainable World* (San Francisco: Sierra Club Books, 2005), 5.
17. Ibid.
18. Dr. Seuss, *The Lorax (*New York: Random House, Inc., 1971), 10.
19. Ibid., 58.
20. Kathleen Dean Moore and Michael P. Nelson, "Moving Toward a Global Moral Consensus on Environmental Action," in *State of the World 2013: Is Sustainability Still Possible* (Washington, D.C.: Worldwatch Institute, 2013), 226.
21. Richard Beach, Jeff Share, and Allen Webb, *Teaching Climate Change to Adolescents: Reading, Writing, and Making a Difference* (New York: Routledge, 2017), 48.
22. Ibid., 52.
23. David Ehrenfeld, *Becoming Good Ancestors: How We Balance Nature, Community, and Technology* (New York: Oxford University Press, Inc., 2009), 22 –3.
24. Ibid., 224–27.
25. Ibid., 226.
26. Tina Lynn Evans, *Occupy Education: Living and Learning Sustainability* (New York: Peter Lang Publishing, Inc., 2012), 238.
27. Paul Shepard, "Nature and Madness," in *Ecopsychology: Restoring the Earth Healing the Mind*, ed. Theodore Roszak, Mary E. Gomes, Allen D. Kanner (San Francisco: Sierra Club Books, 1995), 21.
28. Ibid., 36.
29. Ibid., 39.
30. Heather L. Reynolds, "Overview," in *Teaching Environmental Literacy: Across Campus and Across the Curriculum*, ed. Heather L. Reynolds and Eduardo S. Brondizio and Jennifer Meta Robinson (Bloomington: Indiana University Press, 2010), 18–19.
31. Scott Russell Sanders, "A Sense of Place," in *Teaching Environmental Literacy: Across Campus and Across the Curriculum,* ed. Heather L. Reynolds and Eduardo S. Brondizio and Jennifer Meta Robinson (Bloomington: Indiana University Press, 2010), 76.
32. *The Hobbit*: *An Unexpected Journey*, directed by Peter Jackson (2012, Warner Bros. Pictures, 2013), DVD.

33. Delafruz R. Williams and Sarah Taylor, "From Margin to Center: Initiation and Development of an Environmental School from the Ground Up," in *Ecological Education in Action: On Weaving Education, Culture, and the Environment*, ed. Gregory A. Smith and Dilafrus R. Williams (Albany: State University of New York Press, 1999), 85.
34. Ibid.
35. Ibid.
36. Ibid.
37. Ibid.
38. McNamee, Thomas. *Alice Waters and Chez Panisse: The Romantic, Impractical, Often Eccentric, Ultimately Brilliant Making of a Food Revolution.* (New York: Penguin Group, 2007), xiv.
39. Delafruz R. Williams and Sarah Taylor, "From Margin to Center: Initiation and Development of an Environmental School from the Ground Up," in *Ecological Education in Action: On Weaving Education, Culture, and the Environment*, ed. Gregory A. Smith and Dilafrus R. Williams (Albany: State University of New York Press, 1999), 87.
40. Richard Louv, *The Last Child in the Woods* (Chapel Hill: Algonquin Books of Chapel Hill, 2005), 19.
41. Ibid.
42. Ibid., 34.
43. Ibid., 35.
44. Ibid., 43.
45. Ibid., 36.
46. Olmert, Meg Daley. *Made for Each Other: The Biology of the Human-Animal Bond.* (Cambridge: Da Capo Press, 2009), 153.
47. Ibid.
48. Rebecca Young, "McLovin: Our Food and What We Can Do About It," *Ragazine, The Global Online Magazine of Arts, Information, and Entertainment,* November–December 2014, 10, no. 6.
49. Olmert, 153.
50. Wendell Berry, *The Unsettling of America*, 129.
51. Homer, *The Odyssey*, quoted in Berry, *The Unsettling of America*, 128.
52. Berry, 129.
53. Winona LaDuke, *One Thing to Do About Food,* ed. Alice Waters, *The Nation,* September 11, 2006, 18.
54. Sheila Watt-Cloutier, "The Inuit Right to Culture Based on Ice and Snow" in *Moral Ground: Ethical Action for a Planet in Peril*, ed. Kathleen Dean Moore and Michael P. Nelson (San Antonio: Trinity University press, 2010), 27.
55. Ibid., 29.
56. Thomas Berry, *The Great Work: Our Way into the Future* (New York: Three Rivers Press, 1999), 177.
57. Ibid., 23.
58. Malcolm Margolin, "Indian Pedagogy: A Look at Traditional California Indian Teaching Techniques," in *Ecological Literacy: Educating Our Children for a Sustainable World*, ed. Michael K. Stone and Zenobia Barlow (San Francisco: Sierra Club Books, 2005), 76.
59. Bill McKibben, *The End of Nature*, Introduction, XV.
60. Bill McKibben, *The End of Nature*, 7.
61. Bill McKibben, *Enough: Staying Human in an Engineered Age* (New York: 2003), 8-10.
62. Ibid., 112-113.
63. Ibid., 112.
64. Ibid., 113.
65. Ibid., 213.
66. Ibid., 213-214.
67. Rachel Carson, *Silent Spring* (New York: First Mariner Books edition, 2002), 3.

Afterword

David W. Orr

We assumed that better facts would make for better behavior and sometimes they do, but humans are seldom so simple. Facts do matter, of course, but not so much without a compelling narrative that places them into a recognizable context. This is where literature sometimes merges with history. When Abraham Lincoln, for example, met Harriet Beecher Stowe the author of *Uncle Tom's Cabin* (1852), he remarked "so you're the little woman who wrote the book that started this great war." The sheer weight of the moral and human facts did not rouse the nation to respond much to the evils of slavery until Stowe put them into a plausible and powerful literary context with faces, names, and circumstances and so bent the arc of history.

Writers do not always rise to the occasion however. Novelist Amitav Ghosh believes that future generations will see "ours [as] a time when most forms of art and literature were drawn into the modes of concealment that prevented people from recognizing the realities of their plight . . . this era, which so congratulates itself on its self-awareness, will come to be known as the time of the Great Derangement."[1]

His point is telling. Until quite recently, climate change did not feature prominently in contemporary literature. That may explain part of our national lethargy in confronting the consequences of our addiction to fossil fuels. The subject is too big and abstract for many to comprehend. And most of us have a good thing going, living lives of cheap fossil fuel–powered mobility and convenience and wishing it permanent. The blame, in other words, is widely shared from the many culpable by dependence on fossil fuels to the corporations that profit indecently. Television news and much of the print media also downplayed and often denied the evidence accumulating all around the world. They rendered it piecemeal as entertainment for higher ratings not as a pattern of dereliction leading to the edge of an abyss.

Things, however, are heating up—both literally and figuratively. Carbon dioxide continues to accumulate in the atmosphere where it will remain for a long time to come. Records for heat, rainfall, drought, fires, floods, wind speed, and sea-level rise continue to fall at a record-breaking pace. Oceans are becoming more acidic and species are disappearing. We are now at the dawn of a new geologic era or, if you prefer, of a different planet. The truth of the matter is that the Earth that nurtured *Homo sapiens* for the past two-hundred thousand years is becoming a less hospitable and more capricious planet. This is difficult for us to comprehend because of the sheer enormity of the fact and because we—as visual creatures—have few images of what that world might be like. Our empathy and moral concerns, accordingly, peter out at about the second generation beyond our own lives.

We do know, however, our ecological shadow will fall heavily on our progeny. We know too that planetary-scale changes will last for a long while. The years ahead will be what James Howard Kunstler has called "the long emergency," and its duration and severity will challenge humans in unimaginable ways. The best that we can hope for, in Bill McKibben's words is to "cap off the worst that could happen." Others, are slightly more optimistic placing their bets on our capacity to make a rapid worldwide transition to renewable energy and also deploy the technology and make land use changes sufficient to suck enough carbon out of the atmosphere fast enough to stabilize climate at some level below that which would destroy civilization.

Writers like Margaret Atwood, Cormac McCarthy, Kim Stanley Robinson and others have scouted possible futures of the industrial world and various human responses to collapsing ecologies, institutions, and social norms. Atwood's description in *Oryx and Crake* of the ruins of tall buildings far out in rising oceans, for example, is a haunting image of a future with drowned coastal cities, a marked contrast to the happy talk about urban adaptation. As I write, the news media is full of pictures of Houston and Texas coastal cities flooding after hurricane Harvey in late August 2017. News reports from South Asia at the same time report even worse flooding that has taken over a thousand lives and ruined villages and cities.

Even at the beginning of the long emergency, many of us are suffering catastrophe fatigue. Writers of all kinds can help us keep our perspective and sanity through the transition to a world better than that in prospect. What kind of help do we need? Among other possibilities, I single out the kind of literature and nonfiction writing that nourishes our ecological imagination and our capacity for empathy—traits that place us individually and collectively in a wider and deeper context. Climate change, for example, is a symptom of much larger systemic problems. I think the same is true of the increasing hard-heartedness, violence, and militarization evident in much of the world. Most of us do not see ourselves as whole people implicated and reflected in all of the surrounding patterns of dis-ease and derangement. I am inclined to attribute much of this to the philosophers, psychologists, technologists, media manipulators, corporate titans, and economists who built their castles on the sands of narrow self-interest, devil take the hindmost. Accordingly, we celebrate our uniqueness as individuals and defend our prickly notions of individual rights to the exclusion of community. Increasingly we seek safety in walls and weapons, not in safe and equitable communities. We are a nation of suckers, caught in a web of lies and half-truths about who we are and what we might become. A larger truth, however, is that we live by the mercy and beneficence of people and forces that we cannot comprehend. No one is an island detached from the mainland of history, culture, genetics, family, the mercy of strangers, ecological gifts, and mysteries untold.

Words matter. Writers living and writing in Truth are the truest patriots of the human nation. The truth in the daily headlines is, and always has been, dire. The larger Truth, however, is that within all of us are better angels ready and able to transform our nature and hence our prospects in a world still rich in possibilities. Literature and education proper for our time would not soft-pedal our situation, but it would stress the necessity for an improved and chastened human nature and lift our sight to a farther horizon.

NOTE

1. Amitav Ghosh, *The Great Derangement*. Chicago: University of Chicago Press, 2016, p. 11.

Bibliography

Ableman, Michael. "Raising Whole Children Is Like Raising Good Food: Beyond Factory Farming and Factory Schooling." In *Ecological Literacy: Educating Our Children for a Sustainable World*, 175–183. Edited by Michael K Stone and Zenobia Barlow. San Francisco: Sierra Club Books, 2005.

Acharya, Sourya. "Mirror neurons: Enigma of the Metaphysical Modular Brain." *Journal of Natural Science, Biology, and Medicine* 3, no. 2 (2012 Jul–Dec): 118–124. doi: 10.4103/0976-9668.101878.

Anaya, Rudolfo. *Bless Me, Ultima*. New York: Warner Books, Inc., 1972.

Atwood, Margaret. *Oryx and Crake*. New York: Random House, 2003.

———. *The Year of the Flood.* New York: Random House, 2009.

———. *In Other Worlds: SF and the Human Imagination*. New York: Anchor Books, 2011.

———. *The Handmaid's Tale and Oryx and Crake "In Context"* . *PMLA* vol. 119, no. 3, Special Topic: Science Fiction and Literary Studies: The Next Millennium (May 2004), pp. 513–517.

Baggett, David. and Shawn E. Klein. *Harry Potter and Philosophy: If Aristotle Ran Hogwarts*. In *Popular Culture and Philosophy, Volume 9*. Edited by William Irwin. Chicago and LaSalle: Open Court, 2004.

Beach, Richard, Jeff Share and Allen Webb. *Teaching Climate Change to Adolescents: Reading, Writing, and Making a Difference.* New York: Routledge, 2017.

Bekoff, Marc. *The Emotional Lives of Animals: A Leading Scientist Explores Animal Joy, Sorrow, and Empathy—and Why They Matter*. Novato: New World Library, 2007.

Berry, Thomas. *The Great Work: Our Way into the Future.* New York: Three Rivers Press, 1999.

Berry, Wendell. *The Unsettling of America.* San Francisco: Sierra Club Books, 1986.

———. "Preserving Wildness." In *American Earth: Environmental Writing Since Thoreau*, 516–530. Edited by Bill McKibben. New York: Literary Classics of the United States, 2008.

Board on Agriculture and Natural Resources. *Impact of Genetically Engineered Crops on Farm Sustainability in the United States.* The National Academies Press. 2011.

Bonca, Cornel. "Don DeLillo's White Noise: The Natural Language of the Species," College Literature 23, no. 2 (June 1996), 25–44.

Bowers, C.A. "Changing the Dominant Cultural Perspective in Education." In *Ecological Education in Action: On Weaving Education, Culture, and the Environment,* 161–178. Edited by Gregory A. Smith and Dilafruz R. Williams. New York: State University of New York Press, 1999.

Brin, David. *The Transparent Society: Will Technology Force Us to Choose Between Privacy and Freedom?* Cambridge: Perseus Books Group, 1998.

———. *Earth*. New York: Bantom Books, 1990.

Brown, Peter. *The Curious Garden*. New York: Hachette Book Group, 2009.

Buell, Lawrence. *The Future of Environmental Criticism: Environmental Crisis and Literary Imagination*. Malden: Blackwell Publishing, 2005.

Bugg, John. "'Master of their language': Education and Exile in Mary Shelley's Frankenstein." *Huntington Library Quarterly* 68, no. 4 (December 2005), 655–666. http://www.jstor.org/stable/10.1525/hlq.2005.68.4.655.

Butler, Ocavia. *Wild Seed*: *Seed to Harvest*. New York: Grand Central Publishing, 2007.

Callicott, J. Baird. "Thinking Climate Change Like a Planet: Notes from an Environmental Professor" in *Teaching Climate Change in the Humanities* (New York: Routledge, 2017), 81–82.

California Education and the Environment Initiative.

Carroll, Noël. "The Wheel of Virtue: Art, Literature, and Moral Knowledge." *The Journal of Aesthetics and Art Criticism* 60, no. 1 (Winter 2002): 3–26. http://www.jstor.org/stable/1519970.

Carson, Rachel. *Silent Spring*. New York, NY: First Mariner Books edition, 2002.

Chapin III, F. Stuart. "The Planet Is Shouting but Nobody Listens." In *Moral Ground: Ethical Action for a Planet in Peril*, 25–30. Edited by Kathleen Dean Moore and Michael P. Nelson. San Antonio: Trinity University Press, 2010.

Chomsky, Noam. *Chomsky on MisEducation*. Edited by Donaldo Macedo. Lanham: Rowman & Littlefield Publishers, Inc., 2000.

Churchland, Patricia S. *Braintrust: What Neuroscience Tells Us About Morality*. Princeton: Princeton University Press, 2011.

Clark, Timothy. *The Cambridge Introduction to Literature and the Environment*. New York: Cambridge University Press, 2011.

Collin, Robin Morris. "Restoration and Redemption." In *Moral Ground: Ethical Action for a Planet in Peril*, 82–87. Edited by Kathleen Dean Moore and Michael P. Nelson. San Antonio: Trinity University Press, 2010.

Common Core State Standards Initiative. Preparing America's Students for College and Career. 2012. www.corestandards.org.

Damasio, Antonio. *Self Comes to Mind: Constructing the Conscious Brain*. New York: Pantheon Books, 2010.

Davies, Stephen. "Responding Emotionally to Fictions." *The Journal of Aesthetics and Art Criticism* 67, no. 3 (Summer 2009): 269–284. http://www.jstor.org/stable/25622077.

DeLillo, Don. *White Noise*. New York: Viking Penguin Inc., 1985.

DeVries, Scott. "Swallowed: Political Ecology and Environmentalism in the Spanish American 'Novela de la Selva,'" *Hispania* 93, no. 4 (December 2010), 535–546. http://www.jstor.org/stable/257582232.

Diamond, Jared. *Collapse: How Societies Choose to Fail or Succeed*. New York: Penguin Books, 2005.

DiSalvo-Ryan, Dyanne. *City Green*. New York: William Morrow and Company, Inc., 1994.

Dougherty, Charlotte P., Sarah Henricks Holtz, Joseph C. Reinert, Lily Panyacosit, Daniel A. Axelrad, and Tracey J. Woodruff. *Dietary Exposures to Food Contaminants across the United States*. Environmental Research Section A 84, 170–185 (2000).

Dussinger, John A. "Kinship and Guilt in Mary Shelley's "Frankenstein."" *Studies in the Novel* 8, no. 1 (spring 1976):38–55. http://www.jstor.org/stable/29531766.

Duster, Troy and Elizabeth Ransom. "One Thing to Do about Food," 17. Edited by Alice Waters, *The Nation*, September 11, 2006.

Easterlin, Nancy. "Cognitive Ecocriticism: Human Wayfinding, Sociality, and Literary Interpretation." In Zunshine, L. (Ed.), *Introduction to Cognitive Cultural Studies*. (257–274). Baltimore: The Johns Hopkins University Press, 2010.

Ehrenfeld, David. *Becoming Good Ancestors: How We Balance Nature, Community, and Technology*. New York: Oxford University Press, Inc., 2009.

Evans, Tina Lynn. *Occupy Education: Living and Learning Sustainability*. New York: Peter Lang Publishing, Inc., 2012.

Fearing, Jennifer, Robert Martin and Matthew Newman. *The Role of Economists in Animal Law: A Report of the Pew Commission on Industrial Farm Animal Production.* Johns Hopkins Bloomberg School of Public Health, 2009.

Fitzgerald, F. Scott. *The Great Gatsby*. New York: Simon & Schuster, Inc., 1925.

Foer, Jonathan Safran. *Eating Animals*. New York: Little, Brown and Company, 2009.

Foster, Thomas C. *How to Read Literature Like a Professor*. New York: HarperCollins, 2003.

Franzen, Jonathan. *Freedom*. New York: Farar, Straus and Giroux, 2010.

———. "Perchance to Dream." *Harper's Magazine*. 1996. 35–54.

Friedman, Thomas L. *Hot, Flat, and Crowded: Why We Need a Green Revolution—And How It Can Renew America*. New York: Farrar, Straus and Giroux, 2008.

Fulkerson, Lee, dir. *Forks Over Knives*. 2011; New York: Virgil Films and Entertainment, LLC. DVD.

Gaiman, Neil. "Why our future depends on libraries, reading, and daydreaming." *The Guardian*. October 15, 2013. http://www.theguardian.com/us.

Gay, Kathlyn. *Superfood or Superthreat: The Issue of Genetically Engineered Food.* Berkeley Heights: Enslow Publishers, Inc., 2008.

Gerzon, Mark. *American Citizen, Global Citizen: How Expanding Our Identities Makes Us Safer, Stronger, Wiser—And Builds a Better World.* Boulder, CO: Spirit Scope Publishing: 2010.

Ghosh, Amitav. *The Great Derangement: Climate Change and the Unthinkable.* Chicago: The University of Chicago Press, 2016.

Gilderhus, Nancy. "The Art of Storytelling in Leslie Silko's 'Ceremony.'" *The English Journal* 83. No. 2 (Feb. 1994): 70–72.

Gleason, William. "Looking Back to Look Ahead: Climate Change and US Literary History." In *Teaching Climate Change in the Humanities.* New York: Routledge, 2017.

Goodall, Jane. *Harvest for Hope: A Guide to Mindful Eating.* New York: Warner Wellness, 2005.

Gordimer, Nadine. "Witness: The Inward Testimony." *Nobel Centennial Symposia.* www.nobelprize.org.

Gore, Al. *Earth in the Balance: Ecology and the Human Spirit*. New York: Rodale, Inc., 2006.

———. *Our Choice: A Plan to Solve the Climate Crisis.* New York: Rodale Inc., 2009.

Grandin, Temple. *Animals Make Us Human: Creating the Best Life for Animals*. New York: Houghton Mifflin Harcourt, 2009.

Griffin, P.G. "Sound & Fury, Signifying . . . What, Exactly? Notes on the Franzen Wars." *The Common Review* 9, no. 2 (2011):14–18.

Grossman, Lev. "Jonathan Franzen: Great American Novelist." *Time*, (2010).

Guggenheim, Davis, dir. *An Inconvenient Truth*. Hollywood: Paramount Pictures, 2006, DVD.

Hardin, Garrett. "from The Tragedy of the Commons." In *American Earth: Environmental Writing Since Thoreau*. Edited by Bill McKibben, 438–450. New York: Literary Classics of the United States, 2008.

Harraway, Donna J. *When Species Meet*. Minneapolis: University of Minnesota Press, 2008.

Hauter, Wenonah. *Foodopoloy: The Battle Over the Future of Food and Farming in America.* New York: The New Press, 2012.

Hawken, Paul. *Blessed Unrest: How the Largest Social Movement in History Is Restoring Grace, Justice, and Beauty to the World.* New York: Penguin Books, 2007.

Hawkins, Ty. "Assessing the Promise of Jonathan Franzen's First Three Novels: A Rejection of "Refuge." *College Literature* 37, no. 4 (2010):61–87.

Hayden, Mary, Jeff Thompson, and George Walker. *International Education in Practice: Dimensions for National and International Schools.* London and New York: Taylor and Francis Group, 2002.

Heath, Chip and Dan Heath. *Made to Stick: Why Some Ideas Survive and Others Die*. New York: Random House, 2007.

Hensher, Philip. "Family Fallout." *Spectator* 46 (2010).

Hillman, James. "A Psyche the Size of the Earth: A Psychological Foreword." In *Ecopsychology: Restoring the Earth Healing the Mind.* Edited by Theodore Roszak, Mary E. Gomes, and Allen D. Kanner, xvii–xxiii. San Francisco: Sierra Club Books, 1995.

Hive Web Networks. *Play Again*, Ground Productions, accessed March 15, 2017.

Hogan, Patrick Colm. "On Being Moved: Cognition and Emotion in Literature and Film. In *Introduction to Cognitive Cultural Studies*. Edited by Lisa Zunshine. Baltimore: The Johns Hopkins University Press, 2010.

———. "Literary Universals." In *Introduction to Cognitive Cultural Studies*. Edited by Lisa Zunshine, 37–60. Baltimore: The Johns Hopkins University Press, 2010.

Holm, Sharon. "The 'Lie' of the Land: Native Sovereignty, Indian Literary Nationalism, and Early Indigenism in Leslie Marmon Silko's *Ceremony*. *American Indian Quarterly* 32, no. 3 (Summer, 2008), 243-274.

Huang, Hsinya. "When Sea Levels Rise: Writing/Righting Climate Change in Pacific Islanders' Literature" in *Teaching Climate Change in the Humanities* (New York: Routledge, 2017), 263.

Huxley, Aldous. *Brave New World*. New York: Harper and Row, Publishers, Inc., 1932.

Israel, Ron. *Global Citizenship: A Path to Building Identity and Community in a Globalized World*. Lexington: Ron Israel, 2011.

Jackson, Peter, dir. *The Hobbit: An Unexpected Journey*. 2012; Warner Brothers Pictures, 2013. DVD.

———., dir. *The Lord of the Rings: The Fellowship of the Ring*. 2001. Warner Brothers Pictures, DVD.

Keen, Suzanne. "A Theory of Narrative Empathy." *Narrative* 4, no. 3. (October 2006): 207–236. http://www.jstor.org/stable/20107388.

———. *Empathy and the Novel*. Oxford University Press, 2007.

Kennar, Robert, dir. *Food, Inc*. 2008; Montréal: Alliance Films. DVD.

King, Roger J.H. "Narrative, Imagination, and the Search for Intelligibility in Environmental Ethics." *Ethics and the Environment* 4, no. 1 (1999): 23–38. http://www.jstor.org/stable/40338957.

Kingsolver, Barbara. *The Poisonwood Bible*. New York: HarperCollins Publishers, 1998.

———. *Animal, Vegetable, Miracle: A Year of Food Life*. New York: HarperCollins Publishers, 2007.

———. *Flight Behavior*. New York: HarperCollins Publishers, 2012.

Klein, Naomi. *This Changes Everything: Capitalism vs. The Climate*. New York: Simon & Schuster, 2014.

LaDuke, Winona. *One Thing to Do About Food*. Edited by Alice Waters, 18. *The Nation,* September 11, 2006.

Lanza, Dana. "Tapping the Well of Urban Youth Activism: Literacy for Environmental Justice. In *Ecopsychology: Restoring the Earth Healing the Mind*. Edited by Theodore Roszak, 213–226. Mary E. Gomes, and Allen D. Kanner. San Francisco: Sierra Club Books, 2005.

Lappé, Frances Mooor. EcoMind: *Changing the Way We Think, to Create the World We Want*. New York: Small Planet Institute 2011.

LeGuin, Ursula. *The Left Hand of Darkness*. Introduction. New York: Walker and Company, 1969.

———. "The Ones Who Walk Away from Omelas." http://www.rohan.sdsu.edu/faculty/dunnweb/rprnts.omelas.pdf.

Lettero, Carly. "Spray Glue Goes. Maggots Stay." In *Moral Ground: Ethical Action for a Planet in Peril*. Edited by Kathleen Dean Moore and Michael P. Nelson, 103–107. San Antonio: Trinity University Press, 2010.

Lipscomb, Benjamin L Bruxvoort and W. Christopher Stewart. "Magic, Science, and the Ethics of Technology." In *Harry Potter and Philosophy: If Aristotle Ran Hogwarts*, edited by David Baggett and Shawn E. Klein. In *Popular Culture and Philosophy*, 9, edited by William Irwin (Chicago and LaSalle: Open Court, 2004), 77–91.

Louv, Richard. *Last Child in the Woods: Saving Our Children from Nature-Deficit Disorder*. Chapel Hill: Algonquin Books of Chapel Hill, 2005.

Luce-Kapler, Rebecca, Dennis Sumara and Tammy Iftody. "Teaching Ethical Know-How in New Literary Spaces." *Journal of Adolescent & Adult Literacy* 53, no. 7 (April 2010): 536–541. http://www.jstor.org/stable/25653903.

Luke, Timothy W. *Ecocritique: Contesting the Politics of Nature, Economy, and Culture.* Minneapolis: University of Minnesota Press, 1997.

Lyna, Mark. *Six Degrees: Our Future on a Hotter Planet.* Washington, D.C.: National Geographic Society, 2008.

Margolin, Malcolm. "Indian Pedagogy: A Look at Traditional California Indian Teaching Techniques." In *Ecological Literacy: Educating Our Children for a Sustainable World.* Edited by Michael K. Stone and Zenobia Barlow, 67–79. San Francisco: Sierra Club Books, 2005.

Marshall, George. *Don't Even Think About It: Why Our Brains Are Wired to Ignore Climate Change.* New York: Bloomsbury Publishing Plc, 2014.

McKibben, Bill. *The End of Nature*. New York: Random House, 2006.

———. *Enough: Staying Human in an Engineered Age*. New York: Henry Holt and Company, LLC, 2003.

———., ed. *Environmental Writing Since Thoreau: American Earth.* New York: Penguin Group, 2008.

McNamee, Thomas. *Alice Waters and Chez Panisse: The Romantic, Impractical, Often Eccentric, Ultimately Brilliant Making of a Food Revolution.* New York: Penguin Group, 2007.

Mission Statement. Common Core State Standards Initiative. Retrieved from http://www.corestandards.org.

Mission and Strategy. Who we are. Retrieved from http://www.ibo.org/mission.

Moore, Jennie and William E. Rees. "Getting to One-Planet Living" in *State of the World 2013: Is Sustainability Still Possible?* Washington, D.C.: Worldwatch Institute, 2013.

Moore, Kathleen Dean and Michael P. Nelson. "Moving Toward a Global Moral Consensus on Environmental Action." In *State of the World 2013: Is Sustainability Still Possible?* Washington, D.C.: Worldwatch Institute, 2013.

———. "Toward a Global Consensus for Ethical Action," in *Moral Ground: Ethical Action for a Planet in Peril.* San Antonio: Trinity University Press, 2010.

Moylan, Thomas. *Scraps of the Untainted Sky: Science Fiction, Utopia, Dystopia.* Boulder: Westview Press, 2000.

Nestle, Marion. "One Thing to Do about Food." Edited by Alice Waters, 14. *The Nation,* September 11, 2006.

Newman, Leah. "Pacific Salt: A Call for Ocean Protection," *YouTube*. 1 Dec. 2015. Accessed 7 Nov. 2017. https://www.youtube.com/watch?v=T4C2g-PHE4Q.

Niman, Nicolette Hahn. *Righteous Porkchop: Finding a Life and Good Food Beyond Factory Farms*. New York: Harper Collins Publisher, 2009.

Oates, Joyce Carol. "Frankenstein's Fallen Angel." *Critical Inquiry* 10, no. 3. (Mar. 1984): 543–554. http://www.jstor.org/stable/1343306.

Oatley, Keith. "Theory of Mind and Theories of Minds in Literature," in *Theory of Mind and Literature*. Edited by Paula Leverage, Howard Mancing, Richard Schweickert, Jennifer Marston William, 13–26. Purdue University Press, 2011.

Ognibene, Elaine R. "The Missionary Position: Barbara Kingsolver's *The Poisonwood Bible.*" College Literature 30, no. 3 (Summer 2003): 19–36.

Oliver, Jamie. *Jamie's Food Revolution: Rediscover How to Cook Simple, Delicious, Affordable Meals*. New York: Hyperion, 2009.

Olmert, Meg Daley. *Made for Each Other: The Biology of the Human-Animal Bond.* Da Capo Press, 2009.

Orr, David. W. *Ecological Literacy: Education and the Transition to a Postmodern World.* Albany: State University of New York Press, 1992.

———. *Earth in Mind: On Education, Environment, and the Human Prospect*. Washington, DC: Island Press, 2004.

———. *Dangerous Years: Climate Change, the Long Emergency, and the Way Forward* (New Haven: Yale University Press, 2016).

Orwell, George. *The Road to Wigan Pier*. New York: Harcourt Brace Jovanovich, 1958.

Palmer, Alan. "Storyworlds and Groups." In *Introduction to Cognitive Cultural Studies*. Edited by Lisa Zunshine, 176–192. Baltimore: The Johns Hopkins University Press, 2010.

Pennington, Juliet. "The Cafeteria Report," 54–56. Elizabeth Turner, *Vegetarian Times*, Issue 405, vol. 39, no. 8 (September 2013).

Pezzulo, Phaedra C., "Teaching Environmental Communication Through Rhetorical Controversy." In *Teaching Environmental Literacy: Across Campus and Across the Curriculum*. Edited by Heather L. Reynolds and Eduardo S. Brondizio and Jennifer Meta Robinson, 98–107. Bloomington: Indiana University Press, 2010.

Pollan, Michael. "One Thing to Do about Food." Edited by Alice Waters, 16. *The Nation,* September 11, 2006.

———. *The Omnivore's Dilemma: A Natural History of Four Meals*. New York: Penguin Books, 2006.

Quinn, Daniel. *Ishmael*. New York: Bantam Books, 1992.

Ramachandran, V.S. *The Tell-Tale Brain: A Neuroscientist's Quest for What Makes Us Human.* New York: W.W. Norton & Company, 2011.

Rawles, Kate. "A Copernican Revolution in Ethics." In *Moral Ground: Ethical Action for a Planet in Peril*. Edited by Kathleen Dean Moore and Michael P. Nelson, 88–95. San Antonio: Trinity University Press, 2010.

Raworth, Kate. "Defining a Safe and Just Space for Humanity" in *State of the World 2013: Is Sustainability Still Possible?* Washington, D.C.: Worldwatch Institute, 2013.

Reynolds, Heather L. "Overview." In *Teaching Environmental Literacy Across Campus and Across the Curriculum*. Edited by Heather L. Reynolds and Eduardo S. Brondizio and Jennifer Meta Robinson, 17–27. Bloomington: Indiana University Press, 2010.

Rhoads, Robert A. and Katalin Szelényi. *Global Citizenship and the University: Advancing Social Life and Relations in an Interdependent World.* Stanford: Stanford University Press, 2011.

Redford, Robert. "Robert Redford: Give the Reins to the Young." *UNESCO*, 7 Dec. 2015, www.//en.unesco.org/news/robert-redford-give-reins-young. Accessed 10 Jan. 2017.

Riley, Jeannette E. "The Eco-Narrative and the Enthymeme: Form and Engagement in Environmental Writing." *Interdisciplinary Literary Studies* 10, no. 2, Ecopoetics and the Eco-Narrative (Spring 2009), 82–98

Robson, Leo. "I Feel Your Pain: The Difficulty of Being Jonathan Franzen." *New Statesman* (2010): 74.

Roszak, Theodore. "Where Psyche Meets Gaia." In *Ecopsychology: Restoring the Earth Healing the Mind.* Edited by Theodore Roszak, Mary E. Gomes, and Allen D. Kanner, 1–20. San Francisco: Sierra Club Books, 1995.

Salatin, Joel. *The Sheer Ecstasy of Being a Lunatic Farmer*. White River Junction, Vermont: Chelsea Green, 2010

Sancton, Thomas. "Planet of the Year: What on EARTH Are We Doing?" *TIME*, January 2, 1989.

Sanders, Scott Russell. "A Sense of Place." In *Teaching Environmental Literacy: Across Campus and Across the Curriculum.* Edited by Heather L. Reynolds and Eduardo S. Brondizio and Jennifer Meta Robinson, 72–76. Bloomington: Indiana University Press, 2010.

Saunders, Caroline. "The Beginning of Us, Rising—Storytelling for Climate Change." *CONENERGY*, 8 Dec. 2015, conergy.com/de/node/231. Accessed 10 Jan.2017.

Saylan, Charles and, Daniel T Blumstein. *The Failure of Environmental Education and How We Can Fix It.* California: University of California Press, 2011.

Schaefer, Carol Lexa. *The Children's Garden: Growing Food in the City.* Washington: Sasquatch Books, 1994.

Schei, Tonje Hessen, dir. *Play Again*. Ground Productions. http://playagainfilm.com/filmsynopsis/. Accessed March 15, 2017.

Schlosser, Eric. "One Thing to Do about Food." Edited by Alice Waters, 14. *The Nation,* September 11, 2006.

Scott, A.O. "Deep Down, He Wants to Help." *The New York Times*. Movie Review. 2012.

Sepúlveda, Luis. *The Very Old Man Who Read Love Stories*. San Diego: Harcourt Brace, 1995.

Seuss, Dr. *The Lorax.* New York: Random House, Inc., 1971.

Shakespeare, William. *A Midsummer Night's Dream*. In *The Norton Shakespeare*. Edited by Stephen Greenblatt, 814–861. W.W. Norton and Company, Inc., 1997.

Shelley, Mary. *Frankenstein with Related Readings*. New York: Glencoe McGraw-Hill, 2000.
Shepard, Paul. "Nature and Madness." In *Ecopsychology: Restoring the Earth Healing the Mind*. Edited by Theodore Roszak, Mary E. Gomes, and Allen D. Kanner, 21–40. San Francisco: Sierra Club Books, 1995.
Silko, Leslie Marmon. *Ceremony*. New York: Penguin Books, 1977.
Singer, Peter and Jim Mason. *The Ethics of What We Eat: Why Our Food Choice Matter*. USA: Rodale, Inc. 2006.
Siperstein, Stephen, Shane Hall and Stephanie LeMenager. *Teaching Climate Change in the Humanities*. New York: Routledge, 2017.
Srinivas, M. N. "Social Anthropology and Literary Sensibility." *Economic and Political Weekly* 33, no. 39. (Sept. 26–Oct. 2, 1998): 2525–2528. http://www.jstor.org/stable/4407216.
Standage, Tom. *An Edible History of Humanity*. New York: Walker Publishing Company, Inc., 2009.
Stone, Michael K. and Zenobia Barlow. *Ecological Literacy: Educating Our Children for a Sustainable World*. San Francisco: Sierra Club Books, 2005.
Stracher, Cameron. *The Water Wars*. Naperville: Sourcebooks Fire, 2011.
Sutherland, Margaret B. "Education and Empathy." *British Journal of Educational Studies* 34, no. 2. (June 1986): 142–151. http://www.jstor.org/stable/3121323.
Tanenhaus, Sam. "Peace and War." *New York Times Book Review* (19 August 2010): NP. In *Contemporary Literary Criticism*, 309.
Temple, Grandin and Catherine Johnson. *Animals In Translation: Using the Mysteries of Autism to Decode Animal Behavior*. New York: Scribner, 2005).
Terborgh, John. *Requiem for Nature*. Washington, D.C., Island Press, 1999.
Theroux, Paul. *The Mosquito Coast*. London: Penguin Books, 1981.
Veidemanis, Gladys V. ""Frankenstein" in the Classroom." *The English Journal* 75, no. 7 (Nov. 1986): 61–66. http://www.jstor.org/stable/818504.
Veldman, Robin Globus. "Narrating the Environmental Apocalypse: How Imagining the End Facilitates Moral Reasoning Among Environmental Activists," Ethics and the Environment 17, no. 1 (Spring 2012), 1–23. http://www.jstor.org/stable/10.2979/ethicsenviro.17.1.1.
Waters, Alice. "Fast-Food Values and Slow Food Values. In *Ecological Literacy: Educating Our Children for a Sustainable World*. Edited by Michael K Stone and Zenobia Barlow, 49–55. San Francisco: Sierra Club Books, 2005.
———. *Edible schoolyard: A Universal Idea*. San Francisco: Chronicle Books LLC, 2008.
Watt-Cloutier, Sheila. "The Inuit Right to Culture Based on Ice and Snow." In *Moral Ground: Ethical Action for a Planet in Peril*. Edited by Kathleen Dean Moore and Michael P. Nelson, 25–30. San Antonio: Trinity University press, 2010.
Weil, Andrew, M.D. *Why Our Health Matters: A Vision of Medicine That Can Transform Our Future*. Hudson Street Press, 2009.
Weisman. Alan. *The World Without Us*. New York: Picador, 2008.
Williams, Delafruz R, and Sarah Taylor. "From Margin to Center: Initiation and Development of an Environmental School from the Ground Up." In *Ecological Education in Action: On Weaving Education, Culture, and the Environment*. Edited by Gregory A. Smith and Dilafruz R. Williams, 79–102. New York: State University of New York Press, 1999.
Wood, Gillen D'Arcy. "Teaching Teleconnection" in *Teaching Climate Change in the Humanities* (New York: Routledge, 2017), 96.
Zunshine, Lisa. "Theory of Mind and Experimental Representations of Fictional Consciousness." In *Introduction to Cognitive Cultural Studies*. Edited by Lisa Zunshine, 193–213. Baltimore: The Johns Hopkins University Press, 2010.
———. *Why We Read Fiction: Theory of Mind and the Novel*. Ohio State University Press, 2006.

Index

About the Author

Rebecca L. Young is a language and literature content specialist for the Measured Progress and International Baccalaureate Organizations. She also serves as an environmental education consultant with a focus on how interdisciplinary curricula and assessments can shape an ecocritical pedagogy that positions educating young people for a climate changed world at the center of public education's mission. Her work has been published in the Critical Insights series and she contributed to *Teaching Climate Change to Adolescents: Reading, Writing, and Making a Difference*. Young earned her PhD in Environmental Literature Studies at the State University of New York at Binghamton. She grew up in the New York Catskills and currently lives in southern Maine.

www.ingramcontent.com/pod-product-compliance
Ingram Content Group UK Ltd.
Pitfield, Milton Keynes, MK11 3LW, UK
UKHW012041150926
475560UK00001B/124

9 781498 5359